Teaching and Mentoring Writers in the Sciences

Writing Your Journal Article in Twelve Weeks
Wendy Laura Belcher

The Craft of Research
Wayne C. Booth, Gregory G. Colomb, Joseph M. Williams, Joseph Bizup, and William T. FitzGerald

The Craft of Science Writing
Siri Carpenter, editor

The Chicago Guide to Grammar, Usage, and Punctuation
Bryan A. Garner

On Revision
William Germano

Writing Science in Plain English
Anne E. Greene

The Craft of Scientific Communication
Joseph E. Harmon and Alan G. Gross

Write No Matter What
Joli Jensen

How to Write a BA Thesis
Charles Lipson

The Chicago Guide to Writing About Multivariate Analysis
Jane E. Miller

The Chicago Guide to Writing About Numbers
Jane E. Miller

The Chicago Guide to Communicating Science
Scott L. Montgomery

Handbook for Science Public Information Officers
W. Matthew Shipman

The Writer's Diet
Helen Sword

A Manual for Writers of Research Papers, Theses, and Dissertations
Kate L. Turabian

A Handbook of Biological Illustration
Frances W. Zweifel

A complete list of series titles is available on the University of Chicago Press website.

Teaching and Mentoring Writers in the Sciences

AN EVIDENCE-BASED APPROACH

Bethann Garramon Merkle
& Stephen B. Heard

THE UNIVERSITY OF CHICAGO PRESS
CHICAGO AND LONDON

The University of Chicago Press, Chicago 60637
The University of Chicago Press, Ltd., London

Published 2025
Printed in the United States of America

34 33 32 31 30 29 28 27 26 25   1 2 3 4 5

ISBN-13: 978-0-226-82932-6 (cloth)
ISBN-13: 978-0-226-84388-9 (paper)
ISBN-13: 978-0-226-84387-2 (ebook)
DOI: https://doi.org/10.7208/chicago/9780226843872.001.0001

Library of Congress Cataloging-in-Publication Data

Names: Merkle, Bethann Garramon author | Heard, Stephen B. author
Title: Teaching and mentoring writers in the sciences : an evidence-based approach / Bethann Garramon Merkle and Stephen B. Heard.
Other titles: Chicago guides to writing, editing, and publishing
Description: Chicago : The University of Chicago Press, 2025. | Series: Chicago guides to writing, editing, and publishing | Includes bibliographical references and index.
Identifiers: LCCN 2025020664 | ISBN 9780226829326 cloth | ISBN 9780226843889 paperback | ISBN 9780226843872 ebook
Subjects: LCSH: Technical writing | Mentoring in science
Classification: LCC T11 .M416 2025 | DDC 808.06/66—dc23/eng/20250716
LC record available at https://lccn.loc.gov/2025020664

♾ This paper meets the requirements of ANSI/NISO Z39.48-1992 (Permanence of Paper).

Authorized Representative for EU General Product Safety Regulation (GPSR) queries: **Easy Access System Europe**—Mustamäe tee 50, 10621 Tallinn, Estonia, gpsr.requests@easproject.com
Any other queries: https://press.uchicago.edu/press/contact.html

For our parents, who taught us the magic and power of books and language

Contents

Introduction

Welcome to *Teaching and Mentoring Writers in the Sciences*. The very fact that you're browsing this introduction suggests that you agree that mentoring and teaching scientific writing is both important and difficult.[1] Before we launch into our advice and the evidence that supports it, we'd like to take a few pages to outline the task at hand and to say a bit about who we are—and who we think you might be.

What are you going to get in this book?

This book offers you evidence-based recommendations and opinions about how to help students write. Some of our recommendations may surprise you; some you may not quite believe; and some may make you a little uncomfortable. But we know you're here because you want to help students write. You'll be able to do that better if you engage with our recommendations and if, when you feel challenged by what we suggest, you dig into the evidence behind them. If we can start by agreeing on that, we think this will work.

Some things we believe

There's a whole book in front of you, but here are the first things we believe about writing, and about helping developing writers.

1. Teaching writing is part of our jobs as scientists, particularly but not only in academia.
2. Writing should be taught throughout science curricula (in disciplinary "content" courses, not just writing courses), at all degree levels.

3. Writing is a craft, not an end product. That craft rests on a learned set of processes and skills, and thus framing writing as talent is counterproductive.
4. Writing is best taught as a set of transferable skills and processes.[2] Developing writers should learn to recognize conventions and expectations across genres and disciplines, conceptualize writing as both *thinking work* and *communicative work*, and become versed in discussion of goals and techniques for writing for specific audiences.
5. Writing is emotional. When developing writers are asked to pretend otherwise, it gets in the way of productive writing habits and of appreciating (even enjoying) the craft.
6. There are few true rules in writing; most "rules" are people's own personal preferences and quirks. Therefore, emphasizing rules misleads developing writers and is counterproductive. Related: line editing (focusing on grammar, punctuation, word choice, and the like) usually isn't helpful writing instruction; it *is* a great way to waste lots of time as a writing mentor.
7. Scholarship and evidence from fields like writing studies, rhetoric and composition, and inclusive pedagogy can greatly improve the teaching and practice of writing in the sciences—and make it easier, too.
8. Mentors who prioritize writing support should be recruited, valued, and rewarded.

These tenets are the foundation of our book. Our stance is rooted in evidence-based approaches to teaching and mentoring that result in better writing. But many scientists don't know about them or use them—partly because the literature supporting these approaches is difficult to access. We think that's a shame and we'd like to help, for your sake and for the sake of the developing writers you work with. If that sounds good to you, our prescription is simple: keep reading.

What's the task at hand?

Many scientists make two discoveries about writing. First, when they begin to write grants, scientific papers, reports, and all the rest, they discover that writing is *hard*.[3] Second, a little further along in their careers, scientists start trying to help more junior scientists write those same things—and they discover that teaching and mentoring writing is *even harder*.

What kind of teaching or mentoring are we talking about? Some scientists teach writing to first-year undergraduates, perhaps in introductory labs or in a STEM (science, technology, engineering, and mathematics) analogue to freshman composition classes. Others teach scientific writing to seniors in the classroom. Still other scientists find themselves in one-on-one (or at least, one-on-few) mentoring situations, working with honors undergraduates, graduate students, postdocs, and other early-career writers. Of course, many scientists do more than one of these things (we certainly have!). So, which of these situations is this book for—and how many other books do you *also* need to read to help students write?

There's good news: in important ways, these teaching and mentoring situations are not actually as different as they seem. Most of the advice we offer applies (or at least can be easily adapted) to all of them. So, we suggest that you may only need this book, at least for now. There are good reasons; for instance, for coordinating and scaffolding writing instruction across entire undergraduate curricula;[4] and approaches for efficient assessment of undergraduate lab reports or papers can be harnessed for efficient feedback on drafts of graduate thesis chapters.[5] This is not to say there aren't important differences in how students are prepared or how they approach writing; but you needn't fear that you must learn and deploy an entirely different set of teaching or mentoring techniques to address each student and each situation.

What *is* this writing that we're talking about? While our focus is on writing in the sciences, we see that as a broader realm than you might expect. We'll certainly talk about "scientific writing," by which most folks mean technical writing such as research papers, grant proposals, monographs, and "student" genres like lab reports and project proposals that are intended to model professional writing. However, scientists write more than that. There are policy briefs, op-eds, essays, blog posts, and popular books, as just a few examples. Increasingly, "science writing" is used as an umbrella that encompasses all these genres.[6] We recommend having students engage with multiple genres of science writing.

We should make one small clarification before going on. This is a book about teaching and mentoring writing, but it isn't a book about *assessing (or grading)* writing. We'll make occasional observations about that, but most of the issues around assessment are general (that is, not just about writing) and could fill a whole book of their own.[7]

Why this book?

Most scientific-writing books and most books about public-facing science writing are primarily "self-help" books that provide advice to a writer about how to construct or pursue certain types of writing.[8] There are also many books for instructors of writing, but nearly all of those focus on people who teach in English (or allied) departments. Despite our efforts to find something that bridged the gap, we couldn't find a single book written for you: mentors and instructors of developing writers working to produce writing about science. So, we wrote this one.

In a short book, we can't provide detailed, genre-specific advice about every scientific-writing task, or detailed guidance specific to every scientific field. (You'll notice that, since we're both ecologists, many of our examples come from that field. You can easily translate or find equivalent examples for your own field.) To reckon with the breadth of STEM fields and genres, we emphasize building writing skills that apply across fields and genres, over a student's entire career. It's true that in teaching science students you can't ignore scientific writing's particular conventions. But you also shouldn't think of teaching scientific writing, or even science writing, as something cordoned off from the teaching of writing more generally. Students who learn scientific writing need to apply their skills to many kinds of writing throughout their careers, and good mentoring helps that happen. Similarly, on the mentoring side, you can learn to teach scientific writing more easily and more effectively from the scholarship of teaching *any* kind of writing. That means there's much to be learned from the disciplines of writing studies and rhetoric and composition, from the close reading and response practices of the humanities writ large, and from the extensive literature on teaching in those disciplines. But most scientists don't read that literature, or they find it challenging if they do. So, we offer you a shortcut: that scholarship underlies much of the advice we offer in this book.

How to use this book

Remember the Choose Your Own Adventure books? In a sense, you're holding one. We won't ask you to navigate your way through a quest for the Abominable Snowman (that one was Steve's favorite), but we will offer you some choices in how you engage with our advice.

You have options about how you move through this book. We've tried to write it so you can dip into the chapters you need when you need them, rather than attempting to read cover to cover. If you're quite new to mentoring developing writers, or you're looking for a primer on the basics, we suggest you start by reading chapters 1–5 and the afterword, in that order. If you've mentored writers for a while, you may be more interested in optimizing how you run your courses, lab, or program. In this case, you might want to start with chapters 4–6 and 10.

Concepts and actionable advice

Regardless of where you start in this book, you'll notice that we've deliberately woven together the conceptual and applied aspects of mentoring developing writers. We could have separated the book into theory and practice sections, but we're acutely aware that most folks looking for this advice already don't have time to look for and digest the literature informing the advice. So, we're putting it all together so you can see both why and how at once, at your fingertips.

Endnotes

You've probably already noticed that this book is sprinkled liberally with endnotes. Should you read them? It depends.

We've tried to thread a bit of a metaphorical needle with the relationship between our main text and the endnotes. We aimed to keep the main text compact and engaging, packed with the kind of actionable advice we think you're probably looking for. If you're mostly here to find out "What should I do?," and if you're willing to trust that there's evidence behind our advice, you can just use the endnote pages to press wildflowers. (A fine use for any book!)

On the other hand, perhaps you want more. Perhaps you're interested in the scholarship that backs up our assertions; perhaps you'd like to be connected with more technical literature so you can learn in more detail; or perhaps there's a particular topic on which you'd like a deeper dive. If so, the endnotes are for you. We won't pretend that they provide an exhaustive review of the scholarship on mentoring writers, but they can get you started: we've pointed to many of the major (and most approachable) sources. You might even find it useful to read just the main text at first, then dip into the endnotes on subsequent (re)readings. You can choose your own adventure.

Being realistic

Whatever your reading plan, keep this in mind: we don't expect you (or anybody!) to reinvent your mentoring practice overnight. Just as it's challenging for students to learn to write, it's challenging for mentors who learn new evidence about effective instruction to incorporate that into their instructing practices. And, of course, there's complexity, disagreement, and even contradiction in the scholarship of fields like writing studies, rhetoric and composition, and pedagogy—just as there is in your own field, or any other.[9]

Furthermore, not everything we suggest here will work for every mentor. Your circumstances (institution, teaching load, compensation, etc.) and those of your students will have real bearing on what's possible. We don't expect you to change everything at once; instead, try something out and see if it works for you, then dip back into the book and add something else to your toolbox. Like writing itself, mentoring writing is a challenging craft you'll practice, and improve at, throughout your career.

Who are you?

Well, presumably *you* know who you are; but please indulge us a little, because it may help to know a bit about who *we* think you might be. You're surely here because you're involved in some way with teaching or mentoring writers in science (or allied fields). You're convinced writing is important, but you've also found that teaching writing is hard.

Beyond that, we suspect there's a lot of diversity in the room; and what you do and where you work influences the mix of developing writers you're working with and the resources you have available to support you. Some of you are just beginning careers on the "other" side of the student–teacher relationship, while others have been in the trenches for decades but still find teaching writing a time-devouring challenge. You might be a professor in STEM or an instructor in another discipline teaching STEM students, at a large university or a smaller college. You might be a writing center professional, administrator, or member of a department or university committee aiming to enhance students' writing skills across campus. You might work in government, industry, or nonprofits, where writing needs and conventions are quite different from academia. Wherever you are, you likely work in an environment where there's ongoing debate about who should actually be teaching students to write. And, of course, you could be working with writers anywhere in the world. While our own experiences are rooted in Canadian and US modes of discourse and pedagogical frameworks, we recognize that there's geopolitical variation in these things. We're confident, though, that you'll find that the evidence we share is adaptable to the many other settings where writers are mentored.

Finally, we'd like to acknowledge that some of you aren't from the identities that have traditionally structured Western academia (and that academia continues to best support). Perhaps you're a contingent (term-by-term) instructor who cares a lot, but you're still burning out while barely making ends meet. Perhaps you're a non-tenure-track faculty member not wholly enfranchised by administrators or your faculty peers, grappling with less student respect, curricular autonomy, or available time than the tenure-track faculty who often dominate discourse. Perhaps you identify as one or more demographics that have been historically excluded and exploited by Western science. Any or all of these identities may be reasons you find writing instruction even more daunting.[10] We're sure we've missed some aspects of reader diversity here; but in any case, we believe that who you are and what you do matters. We hope to help you do it more easily.

Who are your students?

Perhaps this seems obvious: they're the developing writers you're working to help. But we think it's useful to recognize their diversity, just as we recognized yours in the last section.

The most obvious way developing writers vary is simply the depth of their experience with writing. Some may be first-year university or college students with essentially no experience of scientific writing as a distinct genre; you may encounter them in a small lab course or a vast lecture hall. Others may be third- or fourth-year students who are relatively proficient in formulaic writing such as lab reports, but who still struggle with more complex and authentic writing tasks. Or you may be mentoring undergraduate or graduate researchers, postdoctoral fellows, or other early-career scientists who've begun to write papers or thesis chapters but are discovering it's not as easy as they hoped.

This spectrum of writing experience brings us to a kink in the need to write about all these types of writer relationships: a necessary matter of terminology. In some instances, you're a boss or supervisor, in others an instructor or adviser, and in still others, a collaborator. In a classroom context, we usually talk of "teachers" (or "instructors") and "students." For graduate students and postdocs, we more often use words like "advisor" and "mentor/mentee." But reading "teaching and mentoring" or "students and mentees" over and over again gets old; and in any case, there's enormous overlap between the needs of classroom students and mentees, and no real bright line between them. (After all, if you're teaching the way we're sure you'd like to, that work involves a substantial element of mentorship no matter who it's directed at.) Therefore, in this book we adopt the terms "student" and "mentor," using them rather generally to cover developing writers at all stages and those helping them in any setting. (We hope you'll recognize the sense in which all of us are students long after we stop meeting the strictest definition of the term.) Every time you see "student," "developing writer," or "mentor," we invite you to superimpose whatever words you consider appropriate in your own context.

Of course, career stage isn't the only difference among developing writers. Some writers come to scientific writing from deep experience with writing in other genres—perhaps they're humanities double majors, or perhaps they're pursuing degrees in science after careers (and writing) in other professions. Others will have written only what they've needed to in class or pursued a degree in science out of the misperception that they can thereby avoid writing. Some will be first-generation students while others come from families with long histories in higher education. And, of course, they will come from socioeconomic, cultural, and ability backgrounds across the breadth of human diversity. We'll say more about these factors in chapter 1.

So, as we forge onward with a discussion of writing support for "students," we want to remember: there's a lot of diversity operating in that single word.

Who are we?

We (Bethann and Steve) are the authors, of course—but we think it might help you to know a bit more. We come from quite different backgrounds in some ways, but we've arrived at similar ideas about mentoring writing—and about our colleagues' need for help with that.

Bethann is Professor of Practice in science communication at the University of Wyoming (UW), in Laramie, Wyoming. She's also director of the UW Science Communication Initiative. She has co-developed curriculum for a first-year writing program and teaches science writing and communication at undergraduate, graduate, and professional levels. She publishes regularly in the peer-reviewed domains that science communication (scicomm) draws from, and trains scientists at UW and beyond in ethical science communication. She was founding editor of the Communicating Science section of the *Bulletin of the Ecological Society of America* and co-hosts the *Meteor Scicomm* podcast. Bethann came to academia following a wide-ranging career that included science editing and journalism plus adult literacy, community development, and conservation education. Much of this roving stemmed from her origins as a first-generation transfer student with no clear idea of a typical route through college to a career. Bethann especially enjoys writing about art–science integration, writing funding proposals, and editing manuscripts destined for peer-reviewed publications. She's also a poet, essayist, potter, and illustrator.

Steve is a professor of biology at the University of New Brunswick, in Fredericton, New Brunswick, Canada. He is the author of *The Scientist's Guide to Writing*, a book of advice for early-career scientific writers. He teaches courses and workshops in scientific writing for undergraduate and graduate students and has mentored undergraduate, graduate, and postdoctoral scientists in research and writing. In contrast to Bethann, he comes from a very traditional kind of STEM training, without much formal attention to teaching, rhetoric, or the scholarship of scientific communication—deficits he's worked hard to overcome! For his part, Steve particularly enjoys writing about natural history and about the history and culture of science, both for scientists and for popular audiences.

Together, we think we have things to offer you. We can speak to the scholarship that underlies recommendations for teaching and mentoring writing, but we're also keenly aware of the reality on the ground—what it's like to mentor real STEM students, given the kind of preparation and time most of us have (or lack). Writing this book for you has also taught us a lot, and we're happy that you're here to join the conversation.

1: Getting on the Same Page

UNDERSTANDING AND COMMUNICATING WITH THE DEVELOPING WRITER

How does diversity in student backgrounds and skill sets present both challenges and opportunities? • How can you free students from reproducing the worst features of the literature they read? • How can you help students understand why we use the writing conventions we do, and how conventions differ from rules? • How can you bring students to expect mentors who discuss writing rather than instructors who correct writing?

A bird's-eye view of some challenges you may face when mentoring developing writers

Nearly every scientist will, at some point, find themselves working with developing writers who struggle with the new-to-them genre of scientific writing (or with related genres of science writing). It might happen to you when you're asked to teach a course in scientific writing. More often, you'll assign writing as part of a course focused elsewhere in your discipline, and then you'll either assess that writing yourself or work with teaching assistants who will. If you're running a research lab, you'll mentor students, postdocs, and other personnel as they develop their writing craft. You may even find yourself mentoring your peers—collaborators, coauthors, and friends. If we're making it sound like mentoring writers permeates everything we do as scientists, that's because we think it does; or at least, we think it ought to.

And yet, many scientists—we include ourselves—find it challenging to support developing writers. Few scientists have formal training in teaching, mentoring, or writing. Those are big hurdles, but they're not the only ones. Many instructors and mentors themselves don't *like* writing.[1] You may no longer clearly remember your own early-career writing struggles, making it harder to relate to students' struggles now. Your students are likely to come from quite different educational and experiential backgrounds than you and may be adept at different sets of skills. (Have you ever watched a child dominate an adult at a video game?) They may also be unreceptive to, or poorly prepared for, writing help because they don't (yet) understand writing as a craft to be learned and practiced.

Writing instruction in primary schools in North America tends to facilitate self-expression and to offer reliable cheerleading for developing writers. However, somewhere between kindergarten and graduate school, most developing writers encounter emotional, structural, or even medical obstacles that render writing a chore or worse. There are staggering and heartbreaking statistics about the far-too-many children who aren't safe at school, don't receive adequate tutelage and support there, and may face intersectional challenges including learning differences,[2] inadequate nutrition, abuse, and biases in and beyond the school environment. These and other issues—compounded by negative experiences with writing instruction and feedback—ensure that our classrooms are increasingly filled with students afraid of or very resistant to writing, even while they know they need to become better writers.

Understanding where students go wrong is easier when we recall where we started ourselves. You, too, were once a new scientific writer (possibly also wary of writing). Most likely, as an undergraduate you were asked to produce a piece of scientific writing, a genre different from anything you'd written before and one with its own history, vocabulary, and conventions for content and style. You may have first been assigned to write a lab report, with the exhortation to "write like a paper from the scientific literature." From that beginning, you went on to write review-style term papers, or perhaps an undergraduate thesis. Eventually, you drafted graduate theses and journal papers or policy briefs, essays, op-eds, and other applied or public science materials. As you did so, you gradually became more familiar with scientific writing and related genres. At some point, you began to mentor others. For most developing writers, this is a lengthy and difficult process, supported by mentors who try their best. But those mentors were never trained to teach writing, so there's a lengthy and difficult process on the instructional side, too. If you recognize yourself here, you're in good company: so does virtually every scientist we know.

All this adds up to both a problem and an opportunity. The problem: if you find it difficult, time-consuming, or frustrating to help students learn to write, perhaps your toolbox for mentoring or teaching writing isn't yet well stocked. Neither are your students' toolboxes for *learning* writing. But there's good news, too. The tools you need exist, and this book offers you the opportunity to find and use them.

Some common counterproductive attitudes about writing—and some advice to change for the better

Every student, of course, is different. Nonetheless, over our years mentoring developing writers, we've identified some common attitudes and positions about writing—held by students or by mentors—that pose challenges to learning and mentoring. We'll survey a few of these perspectives here, along with some approaches to overcoming them. Then, we discuss most of these ideas in depth in the rest of this book. Why? Because understanding what developing writers and their mentors bring to the instructional table can set you up, as a mentor, to make rapid progress in equipping your students to tackle their scientific-writing challenges.

1. RECOGNIZE THE EVER-INCREASING DIVERSITY OF STUDENTS' BACKGROUNDS.

We'll start by emphasizing that "getting on the same page" doesn't mean just bringing developing writers to the page their mentors are on. As more senior writers, it's incumbent upon us to realize that our students aren't younger versions of us. Established scientists have been selected, unfortunately, to represent a limited sliver of humanity;[3] but developing writers are much more interesting than that.

Even though there's still a long way to go, the developing writers you work with are part of a student body that represents the diversity of human culture and experience better than ever before.[4] This diversity offers both challenges and opportunities to writing instruction. Developing writers who speak English as an additional language may need to be connected with resources designed for their situation.[5] Instructors and mentors should also be sensitive to developing writers who belong to communities using a range of "nonstandard" Englishes; they are being now asked to produce scientific writing that complies with rather narrow linguistic conventions.[6] Furthermore, some developing writers may rightly object to the gendered language and classist assumptions of "classic" writing guides. We also have more kinds of differently abled students than ever before, and this affects all dimensions of learning. Finally, students are more neurodiverse than ever (or, at least, neurodiversity is better recognized). Although Bethann and Steve both identify as neurodiverse, we aren't neurodiversity-teaching experts. Nevertheless, we suggest that it's important to consider writing instruction—indeed, all instruction—through a neurodiversity lens.[7]

These examples only scratch the surface of the ways student diversity ought to be considered in teaching and mentoring writing. But seeing diversity of culture and experience as a challenge is seeing only half the picture. Your students' diverse perspectives can be embraced to improve both the teaching of writing and the texts that result, if you collaborate with them respectfully.[8] Discussion of the writing obstacles faced by people of different experiences can be enlightening for all; students may be surprised how often different experiences shed light on common pathways for improvement. Writers of different backgrounds can also produce writing with fascinating insights, arresting styles, or unexpected viewpoints. These assets are a powerful force in science and science communication. You might object that in scientific writing, less room is conventionally afforded for this kind of variation. You would not be wrong; but even so, it can be productive to discuss with developing writers why such strong conventions exist. It's also helpful to explore a related question: when does standardization allow for efficient technical communication, and when does it lead to exclusionary attitudes and outcomes?

2. EMBRACE DIGITAL STRENGTH; REMEDY UNFAMILIARITY WITH THE ANALOGUE.

It's a hackneyed stereotype that young people are glued to their screens but never pick up a book. We urge you to resist a "kids these days" reaction that sees the digital divide as evidence of the decline of modern society. Instead, try to remember that developing writers are likely to have a different blend of scholarly strengths and weaknesses than their mentors.

We often find ourselves awed by the digital abilities of our students. They tend to be superb at discovering new papers via social media, personal online networks, or notifications from Google Scholar (although their approach may lean more heavily on haphazard encounter than on systematic searches of scholarly databases). They swiftly locate online information and tools for research or writing. In contrast, they're often less familiar with hard-copy resources such as reference books, print-journal collections, or archives. Developing writers are often less familiar with the act of walking into a library building, navigating the shelves, and borrowing or copying a paper resource. Their writing, as a result, can shortchange knowledge that's a few years old or not easily found in a quick online search. Fortunately, you have powerful allies as you help developing writers use literature and other resources more broadly: librarians. Your institution's librarians are probably willing—even eager—to become involved with efforts to teach writing. (See chapter 7 for an extended discussion of what librarians have to offer.)

We've found that the digital/analogue contrast extends beyond the obvious matter of searching for information. Developing writers are likely to be so skilled at working with text and graphics digitally that it may not occur to them that there's value in picking up a pencil. They may, for instance, be unfamiliar with the benefits of outlining, editing, or proofreading on paper. When asked to design a table or graph, they may instinctively start to work in Excel or Minitab and find themselves too distracted by the profusion of style options to think carefully about fundamental issues (e.g., what columns to include or what to plot against what). It can be productive, as a result, to assign work with paper and pencil, and to ask students to submit comments on their experience with those analogue tools in addition to the actual writing or graphics they produce.

A clear counterpoint exists here. We can't forget that digital tools have made it possible for students to do things, including find information, that would have been completely inaccessible to many of them even a decade or two ago. Various technologies are also vital for people who need assistive devices to read, write, and otherwise operate in the world. Thus, when possible, we recommend analogue assignments be designed so as not to single out students who need digital accommodations.[9] For example, we caution against laptop/device bans or other hardline stances about when and where technology can be beneficial in learning environments.

3. ADDRESS THE CONFUSION BETWEEN TALENT AND SKILL.

Students often express the belief that they aren't "good at" something: math, sports, writing. It's certainly true that humans have innate abilities. Some people are just better than others at hitting a baseball a long way (to pick an arbitrary yet sometimes lucrative example). But professional baseball players also put in thousands of hours of batting practice to develop their skills, regardless of early aptitude. The same is true for writers. This is an equalizing truth: struggling to write is normal for just about everyone.[10] Fortunately, so is struggling progressively less over the span of a career spent practicing and honing the craft of writing.

The confusion of practiced skill with innate talent is understandable, but also critical. Indeed, in our combined four decades of teaching writing and communication, we've come to see that learning scientific writing is the simpler of two challenges facing developing writers. The harder? Learning to *learn* to write. There's a widespread belief among students that "writing up the results" is just something one does at the end, often at the last minute, as a routine (if difficult) task that isn't worthy of the kind of deliberate practice and development given to the other research skills needed to conduct science. Perhaps this belief is unsurprising—since few university science programs offer dedicated scientific-writing courses, and fewer still require them—but it's wrong. Writing is a suite of skills and processes, all of which can and should be learned and practiced throughout one's career. These dimensions of writing are as integral a part of science as are learning statistics, pipetting, or programming.

The belief that writing is just something one does—an output, not a craft or identity—drives a lot of student behavior. You may have the (fairly well-founded[11]) impression that undergraduate students churn out writing just before it's due, essentially submitting first drafts that don't appear to even have been proofread. You may suspect that graduate students do the same for conference abstracts and grant applications. This isn't just about procrastination; it comes from the belief that writing just happens, with the quality of the result set by talent rather than work. This belief can be baked in during high school or other early training, and it persists because students rarely get mentored through the iterative processes of writing[12]—and because they rarely see role models (mentors and others) doing the iterative work of writing. The taxing work you do to bring a coherent, polished piece of writing to life is typically invisible to students, who don't see your multiple outlines, rejected funding proposals, nasty comments from reviewer 2, or manuscript ideas that never saw the light of day. Even more hidden is your gradual, career-long growth as a writer. Instead, your students see only your *product*: a published piece of writing. The confusion of talent with practice is thus to be expected. But it poses a critical hurdle for students: their belief that they *can't* write (or at least, not easily or competently) can be demotivating or even debilitating.[13]

Fortunately, there's a flip side that you can facilitate and model: all writers can improve. We both remember realizing that we could practice writing and get better at it—that we weren't doomed to struggle forever—and the realization was absolutely transformative for us. It was especially transformative to learn that getting better at writing could be a process of self-expression and complex learning about both content and the craft of writing. That is how writing became something we actually enjoy.[14]

We've seen the same realization make a huge difference for writers we mentor (box 1.1). *But how can students come to understand that they're unpracticed, not untalented?* And that they're allowed to find pleasure or at least satisfaction in scientific writing? To start, there's considerable value in mentors taking an autobiographical tone to help students believe that growth is possible. You might get students' attention by disclosing that you got a D on a second-year term paper but now publish routinely in high-impact journals. You might share the popular writing you do, or that you keep a journal, write a blog, or dabble in poetry. You can also share your actual writing process—for example, by asking undergraduates to edit passages from an early draft of your writing or inviting grad students to comment on the current draft of a full manuscript you're working on. You can describe your writing schedule (when you draft, how you handle finding citations, the time you leave between drafting and polishing). You can talk about how and when you solicit feedback from colleagues or collaborators, and how you know when a text is done. The point is to show that you don't rely on innate ability or lightning bolts of inspiration to pour finished writing directly onto paper (or into a digital file). Students tend to be astonished and relieved to learn that even established writers practice a craft by generating and working through many unpolished drafts.

BOX 1.1

Student comments on a growth mindset

Bethann co-teaches a short course for graduate students on what she calls a "scholarly writing mindset." One goal of this course is for students to understand writing as a learned and practiced craft, rather than an innate talent or just an outcome. (Her co-facilitator frames this as seeing writing as an action [verb] versus a thing [noun].[15])

Here's what two students had to say about their realization of the difference:

> "I became intimately aware of how little time I had to write and how long good writing takes. I had to come to terms with and develop a practice for generative thinking beyond writing specific manuscripts, literature reviews, or course assignments."

> "A mindset shift took place. As I got to hear others' writing journeys, I was able to develop a growth mindset. This was possible when others shared their humanness and vulnerabilities. Both faculty members and students were transparent about their experiences, and I thought that was helpful."

Mentors can also put explicit focus on writing as a set of skills and processes (or, if you prefer, as a craft). Students can be asked to read and discuss literature about ways writers can develop.[16] More directly, they can be asked to reflect on their experience, expectations, and strategies for improvement as writers (see chapters 4, 6, and 10 for detailed suggestions here). This should include times when they've struggled but then achieved their writing goals, or times when they've experienced pleasure in writing (even if it was a long time ago or unrelated to science). The main benefit of asking students about these things may be to provoke the realization that they are things one can, indeed, think about.[17] The objective is not to hold the students' hands, but to provide them space and context in which to practice thinking consciously about productive writing habits.[18]

We can embed this notion of practice and skill-building through careful design of courses and curricula. Writing assignments can be scaffolded within courses (e.g., hand in an outline first, then parts of a rough draft, then more polished full drafts), with relevant instruction, feedback, and student reflection at each stage.[19] Similar scaffolding can build writing skills and habits across courses and years of an undergraduate or graduate program (e.g., focus on Methods and Results in first year, with Introduction and Discussion components added in later years; see chapter 10). Students at all levels can be encouraged to explore resources for learning writing and to report on how they used them—one way to support this is to assign a writing guide for a course, or better still, a degree program (see the appendix for some recommendations and chapter 10 for implementation ideas).

For developing writers in research contexts, the lab group can be harnessed for collaborative skill and mindset development. Members might read a writing guide together, with weekly chapter discussions; or they might write together on a joint manuscript or on separate ones, exchanging drafts for peer feedback. Even an occasional lab meeting devoted to writing processes can pay dividends: members can exchange tips while everyone builds awareness that there *are* productive writing processes that can be practiced, and that they aren't alone in their learning-to-write journey. Celebrating incremental writing progress (an outline, a "shitty first draft,"[20] a bit of positive feedback) can do wonders for morale and motivation. Finally, many labs and graduate programs have journal clubs, in which members read and discuss papers in their research areas. Typically these clubs focus on content, but they can be expanded to focus on writing. For an easy first step to connect this activity to skill development,[21] ask each participant to identify something about the paper's writing that they would avoid in their own work, and also something they would choose to imitate. (See table 6.1 for detailed suggestions about how to embed writing training in a graduate student's progression through a research lab.)

All these interventions have one thing in common: they focus attention on writing as a verb (involving processes and skills one can learn) rather than writing as a noun (a product that appears mysteriously and fully formed). This is the essential distinction that most developing writers need help to understand.

4. COUNTER THE URGE TO SOUND "SCIENCE-Y."

When most students begin to produce scientific writing, they need to learn what the genre is like. Their first experience is often a first-year laboratory course, where they're asked to write lab reports "in the form and style of a scientific paper." This assignment is rarely preceded by much instruction about what that form and style actually are, let alone why our literature has adopted them. Instead, students model their writing after the literature they're exposed to. They may begin with their lab manual, they may be given examples of papers or sections of papers, or they may be asked to search for and read papers related to the week's lab. As they move through a university science program, their engagement with published scientific papers grows, but their explicit instruction in writing often doesn't. (Indeed, where specialized scientific-writing courses exist, they're typically optional courses for students in their final year.)

The key problem is that students are asked to model the existing literature without much discussion of which features should be emulated and which should be avoided. There's plenty of the latter; our literature is replete with turgid, tedious writing, soulless passive voice, unnecessary jargon and complexity, acronyms, and much more.[22] These features are obvious, and they contrast conspicuously with the nontechnical writing students know. It's not surprising, then, that the undesirable features of scientific writing make a piece of writing "sound science-y" to students. It's also not surprising that students deploy those features in their own writing—they want to sound science-y, too. Worse: when their work is assessed, teaching assistants or instructors may reinforce these habits. Sometimes this is deliberate, as when inexperienced graders confuse their knowledge of what our literature is like with the notion that somehow rules require the literature to be like that. More often, reinforcement is probably unconscious: we, too, read a piece of turgid and tedious writing and feel some comfort in its conformity with what's already in our literature. We give such writing a higher grade for having "fit in."

It's difficult to improve writing practice when students (and quite possibly mentors) are caught in this stuffy spiral, with bad writing in our past literature encouraging more—or worse—bad writing now. Since at least some of the developing writers in question will go on to publish and peer-review the future literature, we're digging ourselves into an ever-deeper rut. Disrupting this feedback loop is critical—but it isn't easy.[23] Many developing writers experience frustration and confusion as they struggle to fit their work into the supposed norms of scientific writing. But many struggle again, later, when asked to move away from those same supposed norms.

At a minimum, writing assignments should help students navigate what we in fact mean by "writing like the literature." Getting our assignments to that point requires us to explicitly identify features to emulate and features to avoid, and ideally to facilitate some discussion of why particular features of writing end up in one category or the other. We return to this issue in chapter 5, where we explore what happens when different mentors don't agree on which features of writing are "good" and "bad."

5. ADDRESS STUDENTS' UNFAMILIARITY WITH THE RATIONALE FOR WRITING CONVENTIONS.

Scientific writing has a multitude of conventions that distinguish it from other genres. There's nothing unusual about that; romance writing, instruction-manual writing, and newspaper writing all do, too. Some scientific-writing conventions are widely taught, such as our near-universal adoption of IMRaD (*I*ntroduction, *M*ethods, *R*esults, *a*nd *D*iscussion) structure. Others may be implicit or unconscious, even as we adhere quite strictly to them (for example, the relative formality of scientific prose). Still other conventions are in flux: scientific writing began with widespread use of the active voice, shifted almost universally to passive voice in the early-twentieth century, and is now, fortunately, in the process of shifting back to active.[24]

Developing writers may not understand why our writing uses the conventions it does, and that makes it challenging for them to make good decisions about compliance with them. Some conventions have solid rationales, and writing that deviates becomes less effective. IMRaD, for example, has evolved as an efficient finding system for readers, who can customize their reading to target only the information they need, finding it just where they expect. Other conventions are on shakier ground, and some simply reflect runaway feedback of the "got to sound science-y" kind we addressed above. Developing writers may feel the need to conform to such arbitrary conventions—but it's best if they understand that they're doing so to avoid friction with graders, reviewers, or editors, rather than to ensure the effectiveness of their text. Without understanding why our conventions exist, a developing writer can't move beyond mimicking previous literature to choosing when to push back a little. There's an interesting parallel here with the teaching of statistics—in both cases, a "recipe" or rote approach too often divorces students from the underlying rationale, leading to unthinking application of a few "cookbook" approaches.[25]

The fix for this problem seems obvious: instructors and mentors should explain why we follow our conventions, or at least direct developing writers to explanations.[26] Such explanations are more readily available than we're conditioned to expect: for a long time, scholars have studied our writing conventions, their history, and their rationales. Such work is part of the disciplines of science studies, writing studies, and rhetoric[27] and composition, and we'll point to each frequently throughout this book. Both you and the people you mentor can take advantage of this knowledge, but most developing writers need to be steered to it.

6. OVERCOME THE BELIEF THAT "CORRECT" WRITING IS ALL ABOUT RULES.

Maybe it's the legendary ruler-wielding nun, punishing infractions in high school English by rapping student knuckles; or maybe it's the weird popularity of grammar scolding online or in books.[28] Wherever it comes from, there's a widespread belief that learning to write involves learning the "rules" and then complying with them. Many students have this belief; unfortunately, we've found that so do some writing instructors. The ruler-wielding nun, at least metaphorically, isn't *just* a legend.

The thing is, most of the "rules" that get bandied about are not, in fact, rules at all.[29] They're linguistic conventions of varying strength, but they aren't bright lines beyond which written language becomes ineffective or should be disallowed. If you look carefully, in this book you'll find split infinitives, sentences starting with or ending with prepositions, and a variety of other infelicities that could, in some framings of writing instruction, endanger our knuckles. We push gently at the boundaries of the "rules" because doing so can make writing more effective and more engaging. That's true in a book like this, and it's true in scientific writing more generally—even, we'd argue, in that most convention-bound of forms, the journal article.

This doesn't mean you should tell your students that anything goes. All languages and genres have conventions that allow clear communication between writers and readers; departing too far from them imperils understanding.[30] Even when text that departs from convention is still clear, readers, reviewers, and editors have expectations. They might be wrong to object to an "error," but they'll still object, and if that happens too often, it makes the relationship between writer and reader difficult. And as Bethann frequently tells students, the harder a reader has to work, the less open they are to what you're trying to say. To avoid this kind of conflict, it helps to know when to adhere to expectations, and when to flout them.

Developing writers, then, need two things: to learn the "rules" (conventions), and to learn when and how to "break" them in the service of better writing. The latter is the harder part. You may find it productive to discuss some rules, and then to ask students to write something that violates them—and to explain why they chose to violate that rule, in that instance. Alternatively, you can task them to discover examples of rule breaking in the literature of your field, and to comment on the effectiveness of the passages they find. Both of these activities can be fun, not merely instructive—and you should foster positive emotions around writing whenever you can. To complement this, be cautious when assessing student writing: don't declare something "wrong" unless it genuinely is, and take time to ask why the writer chose to depart from convention (and here we're foreshadowing our next section).

If we can tempt students away from a view of writing as rigid compliance with rules, we can teach them that learning to write well is a bit more complicated—and much, much more interesting.

7. MOVE BEYOND "CORRECTION" TO DISCUSSION AND ITERATION.

Most developing scientific writers have far more experience with their writing being graded—once—than with any other form of feedback. They write an essay, a term paper, an assignment, or a lab report. They hand it in, and it comes back with a score, and some red-pencil corrections, and that's it. This dead-end approach differs in two important ways from how experienced writing mentors—and writers—think about the job. First, the focus is too much on correction rather than discussion. Second, one-and-done assessment doesn't allow for gradual improvement through repeated drafts. No wonder our graduate students later show surprise—and resentment—when asked for a third, or fourth, or eleventh draft of a thesis chapter!

Moving developing scientific writers from correction to discussion can be difficult, because most students are conditioned to expect that there's a single correct answer (in content and in writing). Likewise, students have been trained that an instructor's job is to know what the correct answer is and to judge whether or not a student has given it. This is especially true for students who have chosen the sciences as a refuge from the kind of ambiguity more common in the arts and humanities. But unlike calculation or the periodic table, writing rarely has a single correct answer.

In a course setting, one approach to breaking the expectation of correction is to ask students not just to write something, but to defend their choices about how they wrote it—perhaps even in comparison with an alternative. Another is to avoid "correction" when marking up (we use that term to indicate an activity distinct from assessment) written assignment submissions. Instead, you can pose questions, asking the writer what they intended by a passage or a stylistic choice. This invites the student to reflect on the writing choice they made, and to consider alternatives; it works best as a first step in iterative assessment. Changes from one-and-done assessment can, of course, mean increased instructor effort, so careful design is important. (We'll have much more to say about feedback in chapter 3.)

For students writing in a research environment—such as Honors undergraduates, grad students, or even postdocs and other early-career scientists—explicit discussion between mentor and writer can set expectations for the iterative process of writing, commenting, and revision. So can sharing the process of a piece of the mentor's writing, as we noted earlier. Indeed, our own trainees have sometimes been shocked to discover that we routinely go through a dozen drafts of a manuscript, and that our manuscripts, too, come back to us from collaborators or reviewers with page after page of edits and comments.

Five domains of writer development

To tie all these ideas together, we offer you a framework from the literature on writing development that helps you identify where your students are and what they need. We find this framework especially helpful, as it identifies five domains of development for writers and asserts that writers must have competency in all five areas to be effective at their craft.[31] We think you'll find these domains useful in thinking through (1) what domain knowledge the students you work with lack, and (2) how you can best foster their development in those domains.

The five domains are:

- discourse-community knowledge (understanding norms and traditions of writing in a specific discourse community, such as a field or discipline, interest, or cultural group)
- genre knowledge (e.g., understanding distinctions between lab reports, peer-reviewed articles, and op-eds)
- rhetorical knowledge (e.g., knowing that one field or culture is deferential, another more direct with criticism)

- writing-process knowledge (e.g., recognizing writing as a process involving ideation, iteration, drafts, revision)
- subject-matter knowledge (e.g., familiarity with the content of a course, field, or research study)

In our experience, most writing instruction in the sciences concentrates on genre knowledge and subject-matter knowledge. Sometimes attempts are made to convey writing-process knowledge, but as we've already mentioned, the one-and-done writing of essays, test responses, scholarship applications, and so on is much more common. As for discourse-community knowledge: we're aware of efforts to share that, of course. But, too often, we default to saying, "This is just how scientific writing works," which doesn't actually explain things to developing writers. It doesn't help that we're likely used to talking about expectations in our field in a highly abstract way that relies on years of experience to fill in the gaps. Our students lack this experience, and thus stating these expectations doesn't necessarily equip students to understand them. And finally, we rarely discuss the rhetorical strategies of our own fields with each other,[32] let alone with students. This can lead to students replicating what they read without really understanding how it works; for example, they may cite a host of studies in an Introduction section without ever stating how those studies set up the "so what" that establishes the rationale for their own work.

What we suggest is simple: review your methods for teaching or otherwise mentoring writing. Do your approaches address all five domains? If yes: terrific! Write a blog post, a curriculum, or even a book, and let us know how you do it! We're not being facetious—if you're achieving success with developing science writers in these five domains, our community of writing mentors could learn from you. But if you're not doing all these things yet, we encourage you to keep reading as we share approaches you can use to ensure that you're fostering students' growth in all five domains.

What's to be gained?

Why are we so concerned with this kind of sociocultural analysis of developing writers and their mentors? We hope it's obvious: misunderstandings arising from divergent backgrounds or mismatches in assumptions can make both mentoring and learning writing harder than it needs to be. For developing writers, these misunderstandings can lead to vulnerability and confusion about expectations, and around their apparent failure to meet them. As you've probably observed, such negative experiences lead students to avoid writing, and this can spiral toward despair. The harm this does to students is avoidable,[33] and getting mentors and students on the same page about the issues we've identified here will help. Being on the same page also staves off extra work needed to correct missteps that students don't even know they're making. Finally, both the quality of student writing and the rate at which that quality improves will be limited without mutual understanding between mentor and student. When you're teaching writing in almost any setting, then, it's well worth some investment in meta-discourse so everyone understands how writing works and how developing writers can learn to write better.

2: What Learners Do with Writing—and How to Get Them There

Most science undergraduates will never write scientific papers—so what *will* they write? • How can we teach writing so students can transfer their skills to the genres they'll actually write in? • What are authentic writing tasks, and how do they help students learn? • How does the science of science communication contribute to all this?

Writing in the sciences involves more than peer-reviewed journal articles, lab reports, and essay exam questions. It's actually a bit odd that those three genres get the most attention in coursework, given they aren't what most science professionals will spend their careers writing! In this chapter, we explore many of the other kinds of writing involved in science students' careers and dig into ways we can give developing writers the transferable skills they'll need beyond academia. We consider authentic writing tasks (writing that's not just for the instructor), and we discuss advantages and disadvantages of writing that's posted publicly, or at least distributed to some external reader. We also provide an introduction to the science of science communication (scicomm). That's because we find that many students are highly motivated by scicomm work, being aware of its importance for careers involving communication, engagement, policy, and so on (not to mention saving the world).

Oh, the places they'll write!

Most of the developing writers you work with won't have careers in academia, let alone on the tenure track. So, if we mentor only for academic writing like journal papers and lab reports, our trainees will be unprepared for most of the writing they'll do in their careers. Employers have for decades complained that students at all degree levels are unprepared for professional writing demands in diverse workplaces.[1] This gap in training is unfortunate. Students come to us expecting to learn what they'll need for the next stages of their careers; we should at least try to mentor them in ways that recognize the writing they'll do in careers outside academia.

What kinds of writing are we talking about? In any professional setting, science-trained people will need to be effective at email and texting. Many will field questions from journalists or policymakers looking for more clear-cut "answers" than most science can provide, and others will be responsible for sustained educational or public-outreach efforts. Some science-trained people may work in a consulting context where they complete writing-involved projects such as species inventory and monitoring surveys. Some may write proposals for funding or the awarding of contracts, client and government reports, and marketing and policy briefs. Other people will work in government agencies in regulatory and policy settings, where their major writing tasks may involve articulating regulations, responding to public or journalist requests for information, drafting "action" documents like species management plans, and much more. Some students may pursue additional training to become medical professionals and be required to communicate in writing with other care providers, insurance companies, and patients and their families. Some may become K–12 educators and school administrators and provide feedback, policies, management documents, and the like to children, teens, or adults. Other science-trained professionals may move into the nongovernmental/nonprofit sector, where grant writing, recruitment of program participants, and grant management are major writing tasks. Others still may move beyond science per se, working across the entire scope of available professional possibilities.

Not everyone will become Diana Gabaldon (who moved from a PhD in behavioral ecology to write the Outlander series of historical novels); but nearly everyone will write *something*. And outside their professional lives, the students you're mentoring may write blog posts, Wikipedia entries, or op-eds and letters to the editor. It's a long list (heavily influenced, we admit, by our own backgrounds in the life sciences). But it could have been much longer, and that's really our point.

Some developing writers know about this diversity of writing during their classroom days. Others come to realize it when they move into a career. In box 2.1, we provide reflections from students discussing the places their writing has gone after graduation.

Few writing experiences in the sciences are designed to prepare students for the real world's myriad writing responsibilities. Indeed, our narrow academic focus arguably makes people *worse* at writing in a lot of other pro-

BOX 2.1

Student reflections on writing in a career

As a graduate student, one of the key things I learned from writing proposals and manuscripts is that you are pitching a story. That frame helped me find deep enjoyment in the process. Grant writing was particularly helpful because it borders on fantasy for me, as I get to pretend I will be doing this work in an ideal world with limited obstructions. I have used this approach to write a Forbes column, where I had to translate key points about ocean science in a succinct and digestible way. That frame of mind also helps me to write essays and proposals about why marginalized people and their stories matter, and what a more just scientific enterprise might look like (a future that arguably also borders on fantasy). Most recently, I have been using these skills to write policy briefs, where I've connected emerging science to key issues relevant to managers and legislators.

Priya Shukla, Strategic Earth Consulting, about becoming a writer while a PhD student

After graduating, I used my writing skills (including research and documentation) in unexpected ways. For example, as the manager of a birth center, I applied to the federal Medicaid insurance program so our midwife could extend her services to a wider financial demographic. I persevered through seven iterations of the extensive application process and succeeded in securing the midwife's status as a Medicaid provider. She performed services over the course of the next few months and our claims were reimbursed. Unexpectedly, we received a letter alleging fraud on our application, demanding full repayment and suggesting we prepare for litigation. I meticulously documented why our application was valid. Medicaid accepted my arguments and conceded that our approval had been an internal error on their part. We were disqualified from further participation as a Medicaid provider; however, they retracted their allegations against us. Without *transferable* writing skills, I would not have been prepared to advocate for our business in the face of a major [US] federal insurance program.

Heather Bomb, entrepreneur, about the relevance of writing skills from her bachelor of science degree in natural resource conservation

---

fessional contexts.[2] We probably sound like we're judging as we write this—and we are, a little—but we're also acutely aware of the systemic constraints that have perpetuated this imbalance in instruction. Individual mentors like you shouldn't be expected to transform academic systems singlehandedly. But individuals can take action within the current system; that's why we wrote this book, and it's likely why you're reading it. We'll highlight some of the possibilities and provide an evidence-based foundation for you to make the changes you *can* make, at the levels where you have influence.

What we mean by "transferable" writing skills and why a commitment to them is essential

During their time with you, developing writers should gradually gain confidence and capacity to read and write in the numerous genres required for their careers—*even ones you don't directly instruct*. Even writing a lab report should help a student get better at writing other things; at the very least, you hope they'll get better at writing coherent sentences and arranging them into paragraphs and sections. But we can do better than that. Writing instruction contributes to transferable skills when we avoid thinking of writing as task-based (e.g., write a lab report, write a press release). Think instead of writing as a suite of awarenesses and skills—for example, the awareness that each genre will have organizational principles that readers anticipate, and the skill to seek out and apply the appropriate principles in any genre. This metacognitive approach to writing is what we're talking about when we say that writing can be taught as a set of transferable skills.[3]

We'd be surprised if any readers disagree that transferability should be a primary goal in teaching and mentoring writing. In our experience, though, the distance between intent and accomplishment of this goal is vast. It's not just that most coursework and experiential learning in the sciences focus on *academic* writing. It's also that most mentors were trained in this same context (if they were formally trained at all), and so may themselves have to work hard to master and then articulate the conventions of beyond-academic writing.

So, what's the path forward? How do we teach students to write lab reports in a way that helps them later write compelling op-eds? Sparking student metacognition is the key.[4] If we teach only that lab reports have separate Methods and Results sections, that's likely all our students will know. Instead, we should discuss *why* scientific writing uses that format (it hasn't always), why that format is effective, and how it meets reader expectations.[5] From that base, we can coach developing writers to ask whether this way of writing would be helpful for readers of op-eds, or whether other readers might have different expectations. Students might be asked to write complementary documents—a lab report and an op-ed[6]—along with a reflection on the differences between them and why those differences are there. (See chapters 3 and 4 for further exploration of these ideas.)

We're hinting at something here, in suggesting that pair of complementary documents. To equip developing writers well, you must provide them experience with a broader range of writing tasks than you might be used to assigning or mentoring.

Why authentic writing tasks matter for students

We point to the notion of authenticity in writing tasks elsewhere in this book, but here it plays a central role. When writing tasks are authentic, students' writing projects model (or actually are) the types of real writing projects they'd do in professional settings outside the classroom or outside the academy. While this certainly includes the academic writing done during a mentored experience like undergraduate or graduate research, it can go far beyond that.[7]

Authentic writing tasks matter, and are powerful, for some straightforward reasons:

- **Authentic writing tasks model precisely the kinds of writing students will be expected to do after they graduate.** Most students are more engaged by writing assignments that relate to what they intend to do in their careers.[8] Their motivation and performance are typically enhanced even further when they know they're practicing (or actually doing) writing that will matter to people beyond the classroom.
- **Authentic writing connects students directly to the goals of a writing task in a professional setting.** When students conceptualize what they're doing as writing for the instructor, then their goal is simple and limited: to please that instructor. While we've argued that we can teach students to extrapolate from that to situations outside the classroom, it's more straightforward to make the connection to other readers explicit. This can be especially powerful if the writing actually *can* reach the real world, rather than just modeling that possibility. Steve's entomology students, for instance, once had a blog-post assignment, with the option of having their posts published on the Entomological Society of Canada's public blog. Several took this option, and it seemed to deepen their engagement with the assignment.[9]
- **It's easier to provide students with examples of professional writing from authentic settings than it is to invent assignments and writing examples that lack relevance beyond the classroom.** You may also find providing feedback easier if you can involve science-trained people now working in, for example, industry, environmental consulting, or science journalism. Of course, if you look for such external feedback, be mindful of student privacy, and consider our advice in chapter 7 about recruiting and rewarding help from colleagues.
- **It's also easier to provide constructive feedback on writing that's focused on reaching real people about real matters.** Perhaps you're

skeptical—but with such assignments, developing writers can't gloss over essential content, because their reader hasn't been in the classroom with them. They must take more care with sequencing information, to be sure the reader can follow the argument. And they must integrate the principles of ethical science communication (don't just bury a reader in facts, calibrate for your audience,[10] etc.) in order to write compellingly. If you focus your feedback on these things, you and the student are working toward a shared goal of helping the reader (versus just "fixing" a text, seemingly just to suit *you*).

- **You can help students overcome the downsides of academic-ese.** Think flowery or stuffy prose, overly complex syntax, and exclusionary levels of jargon. Shift your students' focus beyond academia to writing tasks that naturally favor more concrete, engaging writing—and these better habits can improve their academic writing, too.

In table 2.1, we develop three examples of how authentic, nonacademic writing could be facilitated in the classroom (see chapter 4 for more extensive discussion of a range of writing assignments). These are not hypothetical examples. We both (along with many colleagues) frequently use these types of assignments to teach students nonacademic genres of writing (e.g., issue-scoping report,[11] policy brief, podcast script, job application). Our examples are framed as work you might assign in a classroom, but of course we encourage the facilitation of authentic writing experiences for graduate students and other more senior developing writers.[12]

There are also benefits to authentic writing tasks that accrue to you as the mentor. Authentic writing assignments offer genuine opportunities for learning on your part as you explore the realms of writing done beyond the academy. As we've noted elsewhere, we've both become more effective academic writers through exposure to, and practice with, communication *outside* academia. Furthermore, the partnerships necessary to provide students with these experiences may have positive impacts on the science you do and the way you share it. Finally, don't discount the value of a feel-good moment: it can feel great to help prepare developing writers for what they'll be doing next. There are few enough positive feedback loops in academia, so take 'em where you can.

A crash course in the science of science communication (and how that's relevant)

Everything we've been talking about in this chapter relates back to central principles of science communication. So, we're going to connect the dots and explain why. We should point out, by the way, that there's a distinction to be made between *scientific* communication (primarily inside science, including scientific writing) and *science* communication (scicomm; outward facing, intended for everyone). Our focus here is the latter.

Because students are motivated by the importance they see in scicomm, and because it naturally involves a lot of writing, scicomm is a potentially powerful component of writing instruction and of developing writers' growth. It's also a genre that many students will write during their careers (or outside their careers, as an avocation). Bethann has written about scicomm extensively elsewhere,[13] and we've outlined possible content for scicomm courses in chapter 4, so here we'll focus more on the evidence base that informs effective, ethical sharing of science. But it's not really just about scicomm: the same core ideas underpin effective writing for all the types of nonacademic writing we've mentioned in this chapter. They're relevant to most academic writing, too.

You might wonder why "ethical" crops up here. By ethical scicomm we mean scicomm that meaningfully accounts for who should be part of the production, dissemination, and use of science (that's everyone, of course). You can argue (and we'd agree) that this is part of scicomm being effective. But sometimes consideration of efficacy gets reduced to being about the goals of the communicator instead of the needs and experiences of receivers.[14] We think it's useful to include "ethical" to remind us of this.

So, what are those central principles? Here are a few both you and your students will benefit from.

- **People don't make decisions based just on "the facts."** It's ineffective and disrespectful to assume people simply need more information to make "good" (meaning "correct according to you") decisions. This attitude is the "deficit model,"[15] and it's the typical mode of training in science (e.g., memorizing facts from K–12 through undergrad). While it's vital to share knowledge with people, information alone just doesn't do it. Prior beliefs, social pressures, convenience, and the like are much more compelling (even to trained scientists) than are facts and statistics.[16] Helpfully, alternatives to the deficit model exist, and you can find discussions and examples in the peer-reviewed literature and applied settings. These alternatives are called lots of different things, including inquiry-based, interactive, project-based, community participation, and more; in essence, they boil down to dialogue (*talking* with people) and co-production (*working* with people).[17] Developing writers often start at a disadvantage in this context, because they've been trained in an environment that assumes the primacy of scientific information and

TABLE 2.1

Examples of authentic, beyond-academic writing tasks for developing science writers

	Interpersonal writing example: Job-application cover letter	**Policy-oriented example:** Policy brief ("one-pager")	**Popular-science example:** Op-ed or article in local outlet (e.g., radio, newspaper, museum newsletter)
Task description	The student researches current job openings in their field, selects one they're interested in, and develops a cover letter to apply, along with their own résumé or CV. They draft, revise, and format the letter to be competitive.	The student develops a one- to two-page policy brief about a science topic relevant to a policy issue active with a specific governing body (local, regional, or national). They don't just research the science (which they may already be doing for their degree). Rather, they connect their science knowledge to the policy situation, perhaps with an issue-scoping report that addresses the science and the socioeconomic and political circumstances of the situation. Students ideally work to understand who the impacted and influential parties are (and what their priorities are); who ultimately makes a policy decision; and how big a role science may actually play in the decision-making. These threads are then used to create a one-page brief discussing how the selected science topic relates to the policy issue.	The student writes a pitch outlining an article or script about a connection between the science they're studying and their community, aimed at a specific local outlet. They then research, outline, draft, and polish the article or script.

Instructor considerations	Whenever possible, encourage students to target a job they would actually apply for. If they're fabricating application materials, they're veering away from the "reality" that makes authentic writing tasks effective.	Outline standard practices in providing policy briefs to decision-makers (e.g., include contact information and effective images/graphic design, keep it short, and assume this is the only reading they'll do on your perspective about that particular issue).	You will need to brush up on the standards of such genres and provide current, accurate guidance to students.
Possibilities for external involvement[a]	The cover letter, either in draft or final form, may be read by an outside reviewer experienced with the job market. If the student is actively on the job market, they are encouraged to actually submit the letter (based, of course, only on their real accomplishments!).	The brief (or a draft) may be reviewed by someone with experience in the science-policy arena who either writes or receives policy briefs. Students may be encouraged to provide the final version to the actual policy maker they are targeting. (A polished draft can later be useful as a writing sample for job applications.)	Pitches or final texts may be reviewed by a science journalist, experienced op-ed writer, etc., or by an editor who commissions or considers similar work. Ideally, the text is eventually deemed polished enough to be published by the partnering outlet.[b] (A polished draft can later be useful as a writing sample for job applications.)
Possibilities for assessment	You could review any or all of the following: background documents (such as a student-made compilation of job ads, issue-scoping reports, or model op-eds or articles); initial and final drafts of the writing project; a reflection by the student detailing their process, what they learned, and how they would refine the text based on the reviewer's feedback; or a reflection on how this kind of writing compares to previous writing the student has practiced. Emphasize the fit between the writing and its target audience (possibly with a student reflection on what they did to craft that fit). If you had an external reviewer provide comments, decide what weight (if any) to give those comments in assessment, sharing this (and your rationale) with the students.		

a. Consider our advice in chapter 7 about recruiting and rewarding help from colleagues, especially when helping you lies outside their job responsibilities.

b. Bethann has built such partnerships into required writing assignments. For example, undergraduates have written and recorded interpretive scripts connecting energy science to artworks in the campus art museum, and graduate students have written natural history "did you know" scripts for the local public radio station (see "Listen Now!" on the Engage Laramie Science site: https://engagelaramie.science.blog/listen-now/). Steve's entomology blog-post assignment (Heard 2016; or, for example, Giasson 2016) is a simpler equivalent.

rewards mastery (or regurgitation) of that information in large quantities. It's hard, with this kind of training, not to just talk *at* people, inundating them with information rather than keeping in mind what they really need. This is not to suggest that there's no role for providing information when it's wanted and needed! Indeed, we presume that you're here, reading this book, because you want information that we can share with you.

- **All people (including scientists) come to any situation with their own values and biases.** Perhaps surprisingly, in many cases attempting to seem totally neutral can make a scientist seem *less* credible (because most people are aware that everyone has values, priorities, etc.).[18] Unfortunately, much of the way we teach and write about science frames it as neutral and objective, even though the history of science is anything but.[19] We need to help developing writers expand their frame of reference so they can be more credible, ethical, and effective.
- **One-size-fits-all is not an effective approach to scicomm.** Due to shifts in media to a more fragmented, niche-viewpoint system and the proliferation of personalized, algorithm-driven social media and news distribution systems, there's no longer a single "general public" (if there ever was). Additionally, most people are swamped with information;[20] even if some scicomm crosses their view, there's no guarantee they'll engage with it. Thus, it's essential to account for others' values and work to identify shared priorities that help connect scicomm with people.
- **Most people's sources of science information aren't scientists or science organizations.**[21] This means we can't expect people to put weight on what a communicator says simply because they're a scientist. Effective communication relies on building rapport instead of assuming authority. And we have to find and meet people where they are (physically or digitally), because they likely aren't seeking out "science sources."
- **Language matters.** Technical terminology and academic-writing styles have their (important) place in how we train developing writers—but so do more applied and accessible styles. It's helpful to acknowledge that we're mostly trained in the language of science, and then work toward language for *sharing* science.

How do these principles help you teach students to write? Understanding them lets you coach students toward writing that's more effective. Beyond that, they can make your instruction more compelling because they help you justify asking students to do the hard work of writing and revising to meet evidence-based high standards. It also makes the assignments themselves more compelling, because you're genuinely connecting students' academic training to the careers they envision. Furthermore, your assessment can home in on what helps real readers connect with a student's ideas. This framing makes it easier to provide concrete comments about what's working and what requires revision.

If we *don't* connect students to the foundations of effective, ethical scicomm, we inflict upon the world more students conditioned to present science in the same off-putting, elitist manner that has contributed to the current sociopolitical morass. Most of us, and most of our students, would like to help pull society out of that mess! Put another way, students want to make a difference through science. They want to leave your mentorship better equipped to contribute their knowledge to science and society. They'll be much more engaged with writing instruction when they believe that you, and the coursework, are preparing them for these goals.

Are there pitfalls in teaching nonacademic writing?

We're tempted to say no, but you probably wouldn't buy that—nor should you. It's going to be extra work to learn about and facilitate writing experiences outside your academic-writing wheelhouse. In the process, you'll try an assignment that falls flat or that students see as busywork. You may experience resistance from colleagues or administrators who prioritize academic writing and see your expanded efforts as a distraction. You may even experience resistance from students whose academic conditioning has led them to disdain other kinds of writing as unserious.

We've pitched the advantage of working to obtain feedback from other people: professionals and others who do the kind of writing you're teaching. This can be both time-intensive—at least, at the start—and reliant on social capital (see chapter 7). When you ask for people's help, you may be turned down, ignored, or countered with a mutual-aid request. But in our experience, you'll also be enthusiastically told "Yes!" by the many people who agree that authentic writing instruction is worth the time it takes.

There are also issues to consider if you're publishing students' work in a public venue. If it's a venue you control, like a course blog, you'll first need to decide what your standards for publication are. Does every student get published, no matter the quality of their writing or thinking? Beyond that, what are your objectives with publication, and how will you help students embrace them? If students' work will be published in a more formal venue (like museum interpretive scripts or newsletter articles), how much time and feedback will be needed to support students in reaching the writing quality necessary for that venue? What about students who prefer not to have their work public? Finally, what about trolls? The chances are small that someone will go after one of your students in a public arena, but they aren't zero. Depending on the topics and venues involved, it may be essential to discuss this risk with your students. Your students will benefit from discussion and even from assignments that help them think through dealing with positive and negative feedback from the public (especially pertinent for social media assignments).

While we've flagged potential trouble spots, we don't want to leave you worried. These issues are manageable, and dealing with them is worth it. You're working to help developing writers practice and grow in writing genres that will matter to them throughout their lives. We predict that your students will surprise you (in a good way) with their engagement and with the writing they produce.

3: Efficient, Productive Writing Feedback

How can you spend less time and effort on writing feedback while having that feedback help your students more? • What blend of positive and negative feedback is most helpful? • How can you calibrate feedback to the type, stakes, and stage of a particular writing assignment? • Do line edits and detailed comments really help students, or can you facilitate more learning with less-detailed feedback? • How can you coach students to self-edit so they need less of your feedback?

Providing feedback on draft writing is a central and powerful approach to mentoring writing. However, without careful design, feedback can be ineffective and time-intensive. For anyone strapped for time—that is, for all of us—it's a deep challenge to find the time for thorough feedback. Moreover, while some of our colleagues enjoy giving writing feedback, others dread it.

Fortunately, there's good news: there are strategies you can use to provide productive writing feedback efficiently. In this chapter, we share evidence-based approaches that can help you toward that goal. We emphasize the context where feedback is most time-demanding: one instructor for many students. This is the structure most obviously associated with undergraduate courses. However, the same strategies are the foundation for the one-on-one situation that's more typical with research students (graduate, undergraduate honors, etc.) and other mentees, including postdocs and early-career colleagues. We're confident that you can extrapolate from our framing of these tips for the former structure to use them in the latter.

We first discuss some big-picture concepts that can help you find a more productive framework for feedback. Then, we drill into concrete techniques you can use to spend less time providing feedback while building students' confidence and investment in their own writing skills. We're not promising the moon here—supporting developing writers isn't *no* work. However, you can find the work considerably more manageable, and more rewarding, if you apply these techniques.

First, though, let's take a look at why writing instruction is hard. This context will help you identify ways to make feedback more effective and efficient.

A reckoning: Why is writing feedback so hard?

Writing is hard work; but many mentors find *teaching* writing even harder. That's especially true for giving feedback on student drafts. In part, this is because most of us have complicated relationships with writing; and if your relationship is sometimes negative, engaging with the challenging parts of others' writing is daunting. At the same time, most experienced writers find it easy to spot an "error," but much harder to explain clearly to a developing writer what's wrong and how to fix it. On top of all that, many mentors haven't found efficient ways to give feedback. And on top of all *that*, instincts about what feedback would be helpful aren't always reliable. These latter problems stem from the reality that writing in the sciences is disconnected from the study of writing and writing instruction. Today, this research finds its academic home in the disciplines of writing studies and rhetoric and composition. Few scientists, though, access this literature.

Let's start with the big picture. Standard Academic English—the language of nearly all science writing—is a distinct dialect.[1] To train someone to write in this dialect, you need the capacity to spot writing issues and provide clear suggestions for resolving them. But, if you learned English as your first language, you've almost certainly been trained in Standard Academic English through immersion, not direct instruction. As a result, there are things for which you may lack vocabulary. You probably have instincts, for instance, about how you would restate, reorganize, or repunctuate this paragraph. But can you articulate—in plain English, for a developing writer—what each problem is, why it might impede readers, and how it could be resolved? That's what developing writers need and what a lot of mentors struggle with.

Because students need explicit instruction with an eye toward increasing independence, our goal for this chapter is to help you move away from merely "correcting" writing. Instead, we'll guide you to provide a balanced menu of praise and critique, to cultivate your students' sense of responsibility for improving their own writing practice, and to identify a draft's level of development and provide feedback appropriate for that level. What we *won't* do, though, is tell you how to assess students' writing. That's a more universal project than this book can tackle (see introduction), and focusing on assessment or grades can get both mentors and students stuck in "correcting" texts rather than improving the developing writer's abilities.

Progress through positive feedback

What *is* writing, and how should that shape feedback? Most theory aligns: writing is an act of expression that reflects how the writer understands and engages with ideas. In academic and professional settings, we stress that "expression" implies writers sharing their thoughts with readers. Indeed, "writing is an act of confidence" that requires writers to be conscious of their skill and knowledge, for both the content they're writing about and the act of writing.[2] Most mentors agree that a positive interaction—including praise alongside any criticism—is needed if students are to gain confidence and thus take charge of their own development as writers.[3] For example, a mentor can give feedback that appreciates a student's work, while assessing and commenting on it.[4] Research supports our collective hunch that finding at least one aspect of a student's writing that you like is essential.[5] Indeed, negative feedback is widely documented to be demoralizing and demotivating, and positive (low-anxiety or low-fear) writing experiences are crucial for productive skill development. Most of our colleagues would find this uncontroversial, but there's a disconnect: empirically, the feedback mentors provide is rarely positive[6] and when it is positive, it's often vague or superficial.[7]

So, what are you to do? The research, and our experience, suggest that you should:

1. Keep track of how much of your feedback, for each student, is positive versus negative.
2. Work to provide a balance, recognizing that because humans tend to focus on negatives, *more* positive feedback may be needed to maintain a student's confidence.
3. For as many of your positive comments as possible, provide concrete feedback that a writer can build on. For example, rather than "Good title," try "Nice job—your title is catchy and hints at your argument rather than simply stating the topic."
4. Facilitate students reflecting on their past positive and negative experiences of writing, along with discussion and activities to reinforce the positive.[8]

But how do you actually do all this? First, it can be helpful to recognize there are several distinct types of feedback. They fall into two major categories: *formative* feedback focuses on developing or revising a final draft/text while *summative* (or *holistic*) feedback summarizes a draft's strengths and weaknesses (or a writer's development) at the end of a process (a draft, an entire project, or even a unit or course). You're right: summative feedback can contain some formative feedback. Formative feedback can be *corrective*, *directive*, *interactive*, and/or *evaluative*,[9] and each function plays a distinct role in mentoring a developing writer. Being aware of these distinctions can help you plan for the kind of feedback you really mean to provide and thereby keep track of how you're balancing types of feedback.

Quicker and easier feedback: How not to bleed dry

Of course, giving positive feedback seems easier when you don't feel swamped by hours and hours of assessing and "correcting" students' writing.

Alert: grand claim ahead!

If you follow the advice we're about to offer, you'll save actual countable hours you previously sank into commenting on writing. At the same time, you'll find that your students' and mentees' ability to write actually improves. But you should know that in our experience, most instructors and mentors balk at this advice. We share it nonetheless because these tips are the foundation of evidence-based[10] ways to enhance students' writing while not sucking up all your time.

TIP 1: START AT THE BEGINNING: WHAT KIND OF WRITING ASSIGNMENT IS IT?

Not all writing assignments are the same. You wouldn't (or shouldn't) expect a student to approach a weekly, short, ungraded reflection the same way they approach their PhD thesis.[11] You'll want to approach these differently, too, in terms of your pedagogical aims and the extent and kind of feedback you offer.

You can think about differences among writing assignments along at least four axes: *iteration*, *stakes*, *timeline*, and *developmental stage*. Each has implications for how you plan and implement feedback. All parties (mentor, teaching assistants or graders, and students) should have clarity on how a particular writing task is classified. It's worth careful thought about what you, as a mentor, are aiming for, and then clear communication to make sure students understand your intent at each stage.[12]

Iteration

Too often, students submit an assignment, receive an assessment, and that's it. Reasonably, students rarely pay much attention to feedback on one-and-done assignments (at least, beyond checking to see if they should question the assessment). As a result, it's probably not an efficient use of your time to offer detailed feedback on such assignments. If the purpose of an assignment is to measure mastery of content or prompt some reflection, then a one-and-done assignment without feedback on the writing is consistent with your goals. But if your goal is to advance students' writing skills, then we advise replacing one-and-done assignments with iterative ones (students submit a draft, receive feedback, then resubmit at least once). This won't, of course, be a one-to-one replacement, since by definition you'll see an iterative assignment more than once. Effective iterative assignments are carefully designed to scaffold skills, by breaking a major assignment into chunks that incrementally build the draft text *and* the writer's skills. (We address scaffolding assignments in detail in chapters 4 [within a course] and 10 [across a curriculum.])

Stakes

Some writing assignments are more important than others. Low-stakes writing might include practice writing, reflections, outlines, or correspondence that carry little grading weight. Writing with moderate stakes might include a graded lab report or assignment that's one of many in a course, or an early draft of a term paper or thesis chapter. Finally, high stakes writing might include something that's a significant part of a course grade (like an exam or final term paper), or a near-final draft of graduate writing (grant proposal, thesis chapter, written comprehensive exam).

It's wise to make both student and instructor effort commensurate with the stakes. Students will be resentful if they're asked to work hard to produce highly polished writing but their efforts don't translate to substantial course credit. Not only that, they may not be *able* to put in that effort, because they're likely to be dealing with many low-stakes assignments across numerous courses. Adjusting your expectations can provide you with more bandwidth to help students achieve the goals you've set while reducing, or at least focusing, the time you spend on feedback. For example, a low-stakes assignment can be submitted in point form or as a bulleted list, because the ideas are more important than their expression in complete sentences—and now you needn't assess sentence and paragraph structure.

You could use a low-stakes assignment as a sandbox to work on just one or two aspects of writing (for instance, asking students to focus on paragraph-level organization while signaling that other elements of the craft can wait for other assignments or iterations). Such adjustments don't just reduce time and emotional loads on both students and instructor; they increase the likelihood that at least one lesson will be learned. Moderate and high-stakes writing can have higher expectations and commensurately receive more expansive, deeper feedback.

Timeline

Nearly all student writing will be produced to a deadline (even if sometimes a flexible one). But the time allotted for a writer to produce a piece of work can vary enormously: from a few minutes (short-answer exam question) to a few years (PhD thesis). When the timeline is short, the writing involved should always be categorized as first-draft writing, with all the expected "flaws": limited coherence; reliance on terminology to display knowledge; errors of grammar, syntax, and punctuation. In other words, you should never expect timed writing, especially when cognitively demanding, to be polished. It just isn't possible for students to produce polished writing under such circumstances.[13] Communication is key here: the goals of timed, assigned writing should be crystal clear to the writer and repeatedly reinforced for anyone assessing or providing feedback. When there's more time—deadlines are a week, a month, or more into the future—you've given students more time to plan for, draft, and refine writing. You can then hope for work that's more polished,[14] and you can more reasonably see any deficiencies that remain as evidence of skill gaps to be rectified, rather than the inevitable outcome of rushing the job. Finally, longer deadlines open up the valuable option of redesigning a one-and-done assignment to be iterative.

Developmental stage

No matter what your students might assume, writing almost never flows fully formed from the pen. Instead, professional writers (like you!) refine a piece of text through many stages. Writing assignments can and should reflect this.

Students are often surprised at the number of developmental stages a piece of writing passes through. A useful schema is to help them recognize the following stages:

a. Conceptual
   1. Early conceptual: topic selection and rationale
   2. Advanced conceptual: developing understanding of topic background and context
b. Presentation
   1. Early presentation: outlining topic, rationale, background/context to informal readers (peers, instructor, writing center tutors, etc.)
   2. Advanced presentation: refining outline in response to reader feedback; expanding outline into draft text
c. Revision
   1. Early revision: refining organization, structure, and accuracy; adapting content and framing in response to reader feedback
   2. Advanced revision: polishing text at the sentence level

All writing passes through all of these stages—whether or not they're individually executed or evaluated. Formally recognizing them offers advantages for both writers and mentors. Developing writers who don't deliberately consider each stage in the process are likely to introduce, or fail to notice, weaknesses in their draft text. For mentors, identifying the developmental stage of a particular piece of writing lets you help students with (and evaluate) the kind of intellectual work that's going on at that developmental stage.

There's an obvious connection here to the idea of iterative assignments, which take students through stages such as outline, rough draft, and more polished drafts. But even a one-and-done assignment can reinforce that there are stages to every writing effort. You might, for example, have students hand in an outline even if there's no intention of it leading to a full draft. Such an assignment serves to corral their ideas, compel an engagement with notions of structure and flow, and show you the framework they have in their heads—making visible concepts or connections that are present, missing, or misunderstood.

An extra benefit to careful thought about developmental stage is that it focuses your work more efficiently. It's no use correcting punctuation in a topic précis or an outline; that simply isn't the point. In fact, this idea of matching the feedback to the developmental stage of the writing is so powerful that we've made it a tip of its own. On to that now.

TIP 2: GIVE ONLY THEMATIC FEEDBACK ON EARLY DRAFTS.

If you can be sure of one thing, it's that the best writing feedback is serial, with students honing their writing through multiple drafts.[15] If you can be sure of two things, the second is that you don't have time for the first.

The solution begins with scaffolding and carefully tuning your feedback to the developmental stage of the writing. This is likely to have the most dramatic impact on instructor practice for early drafts, but you may be surprised to know what we'd call early: all but the very last stage (c2) in the schema we presented in the last tip. Until the writing is very far advanced, our recommendation is to follow this straightforward process:

1. Skim the writing.
2. Note at least three things the writer did well. Depending on the developmental stage, this could range from including all the relevant ideas (even if they're scattered throughout) to identifying possible counterarguments (even if dismissive of them) to including appropriate references (even when insufficiently integrated).
3. Look for patterns such as redundancy, gaps or jumps in logic, missing concepts or essential citations, and content in inappropriate places. Common issues include:
    a. Content out of IMRaD order (e.g., methods in the Introduction, results in the Discussion).
    b. Jumbled topic/key sentences, paragraphs that need dividing because they're overfull of distinct ideas, and paragraphs that need combining because they elaborate the same idea.
    c. Redundancy either within or between sections.
    d. Central ideas found only at the end of a draft. It's common for students to subconsciously use the writing process to refine their thinking. By the end, they've found the ideas they need. Tell them this is a huge win; it's what the early drafts are for. And then push them to revise by moving those ideas into appropriate—earlier—places in the next draft.
2. Distill these patterns into three to five major points of feedback. For each point, state the pattern you observe in a single sentence.
3. Provide the student with examples from their own writing—you can copy/paste, highlight, cite line numbers, or whatever works for you.
4. Round out the feedback with one or two suggestions (no more!) for how the student should proceed. (Two suggestions total—not two for each feedback point.) You're aiming for high-level patterns the student can recognize, along with manageable advice they can act on.

Then, you await your encounter with the next draft.

Now, how do *more* drafts result in *less* work?[16]

Why it's less work for you

Constraining your feedback to big-picture patterns can get you down from 30 minutes or more—perhaps much more—to about 10 minutes per text. After all, you're skimming, flagging big patterns, and providing two suggestions. (If you're like us, it may take you some practice to get to this brevity.) You'll also find that your students share patterns. Keep a document of template responses, which you'll build up quickly during the first few drafts you comment on. You can copy/paste from there, taking only a moment to customize comments for each student. You can also use your template library to equip TAs to help with feedback, even if they don't have as much experience as you do.

Is it less work for the student?

It isn't *less* work for them—it's *different* work. That's the point. You're shifting the burden of labor from *your* time-consuming, detailed edits to *their* engagement in the writing process. You're helping them to develop awareness of and to manage their own writing patterns. They are now responsible, as active learners, for doing the conceptual work to restructure and refine their writing.

But what if the writing is actually "bad"?

We have colleagues who are on board with this whole streamlined feedback premise until they realize we're asking them to let go of how "good" a student's writing is. They feel it's irresponsible to give a student the impression that their writing is better than it actually is, and that focusing on only a couple of patterns and actions for revision will imply to students that the rest of the text is fine or even "done." Their instinct is to mark up the entire document to be sure the "errors" are flagged, but then assure the student that only certain "errors" need to be corrected at this stage. While this approach is understandable, it's not effective, because it tempts students to focus on accepting corrections to comma splices rather than doing the necessary bigger-picture work of restructuring, including more-effective citations, and the like.[17] You've seen them do it! If you want students to understand that you're asking them to focus on certain things, then emphasize those things and let the rest go until the next round. If you still want to make clear that you are not giving students' overall writing a "pass," you could include in your comments something like the examples in box 3.1.

BOX 3.1

Example comments for clarifying that a student draft isn't polished yet

- While this draft proposal isn't assessed on sentence-level polish, allowing yourself a bit of time to proofread (via reading aloud, printing and hand-editing, or using the free version of an app like Grammarly) can help you catch errors that would interfere with a reader if you were calling this a final draft.
- One comment toward enhancing your work: there are several aspects of the text that assume a lot of insider knowledge on the part of the reader. Ideally, a fully polished draft would provide full context for the situation, involved partners, and other elements of your project "landscape" that are key to understanding the issue. A useful guiding principle for these sorts of documents is to assume the reader is well educated, and interested in but not well informed about the issue.
- While this draft isn't assessed on writing technicalities, another round of proofreading would enhance its clarity and flow. Here are two tips for fine-tuning the writing: (1) Read the text aloud. This can feel a little awkward, but it will help you spot sentences that aren't quite finished (sentence fragments) and sentences that are so long it's hard for the reader to track all the ideas. (2) Use the free version of an app like Grammarly to spot misspellings, typos, and other proofreading issues like misused words. There are some instances of these in your current draft.

Focusing on thematic feedback

What might thematic feedback actually look like? We provide two examples here. The first (table 3.1) is associated with an undergraduate course, with a one-teacher/many-students context. The second (table 3.2) is for a draft from a research student (as it happens, an undergraduate researcher), with a one-mentor/one-mentee context. You'll see, though, that the two share structure and intent.

Mentors of graduate students and postdocs will probably need to make some modifications here. First, the greater complexity of the material means that you'll surely need to spend more than 10 minutes per draft! Second, as teaching shades into co-writing, you're likely to give more detailed and more directed feedback. You may even find yourself contributing to the writing itself as a coauthor instead of, or as well as, a mentor. (We address these considerations in detail in chapters 5 and 6.)

TIP 3: DRILL TOWARD DETAIL THROUGH MULTIPLE DRAFTS.

Your high-level feedback on an early draft will ideally be followed by the receipt of more-mature drafts. That is, thematic feedback is (usually) part of iterative feedback and assessment. While iterative feedback may seem like it will greatly multiply the demands on your time, with careful design it need not. First, you may choose to substitute a single iterative assignment in lieu of several one-and-done assignments. This is most feasible for you as well as for your students if you assign shorter drafts or scaffold the production of longer texts. Second, you can design the iterative feedback carefully to keep the time requirement under control.

The key technique here is a straightforward extrapolation of tip 2. For the first draft, give only thematic feedback, prioritizing issues like topic, organization, and context. For the second draft, you'll focus on organization (again), clarity, and accuracy. In subsequent drafts, as students' writing becomes stronger, you can begin to flag grammatical and syntactical patterns (perhaps using a minimal marking approach; see next tip). In each round, however, you'll keep yourself to a few major feedback points and two suggestions. If you work this way, three rounds of feedback at about 10 minutes each will result in more learning and higher-quality writing than a single 30-minute session spent line editing. And it won't feel like a dead end to you and your students. Granted, there are timing and deadline details to work out here, but you may find it worth tinkering with the syllabus if the product is substantially improved.

Now, a caveat and acknowledgment: You can manage iterative assessment more easily if you don't have to do all of it yourself. The most obvious path here is to build in student peer reviews. Students often give peer review a bad rap, but the literature provides (nuanced) support for peer review when deployed effectively.[18] Intermediate reviews by peers can help students notice shared strengths and challenges and recognize that student growth in writing is possible, while cutting down on the number of drafts you need to handle yourself. There are, of course, issues to watch out for, including students reinforcing each other's "bad habits." (We address student peer review in more detail in chapter 7.)

TIP 4: LET GO OF LINE EDITS.

Most of us are accustomed to line editing and being line edited: every grammatical, syntactical, conceptual, and vocabulary "error" marked, with changes recommended, line by line. That line editing is time-consuming is obvious, and this alone should be reason to relinquish your red pen. But most of our colleagues have a deep-rooted commitment to grammatical and syntactical correctness; you may be among them. So, here's another reason to let go: *line editing doesn't work.*

TABLE 3.1

Example of thematic feedback (public engagement–focused)

Type of project	Communicating about a science topic with a public audience
Assignment context	Draft proposal, including annotated bibliography (15+ references). Proposal is developed over the first half semester, with due dates for individual components. These comments respond to the first full draft (<1,000 words).
Feedback	
(1) Give positive feedback to start.	(1) You have a strong start on your project proposal here. This draft points out a number of interesting aspects of the litter issue affecting your local stream, along with who might be most likely to (a) care and (b) do something about it. Your idea to do a stream cleanup event is likely a productive approach.
(2) Flag missing components—keep your focus on the draft, not the student. Use phrasing such as "the sources," "the text," etc., rather than "you."	(2) However, the complete annotated bibliography wasn't included, and the sources in the annotated bib should be integrated throughout the proposal. The sources are required to inform the way you think about your project and convey that in the proposal. If you'd like to revise and resubmit the proposal with this in mind, I'm willing to review and reassess it. This current draft is also missing a thorough characterization of the target audience. You've been thinking about this, but it's not evident in this draft. An important next step is to think hard about what that audience cares about and how their values might align with cleaner streams. As with the annotated bib, I'm willing to review a revised version.
(3) Suggest revisions (focusing on only two big things).	(3) Here are some big-picture thoughts to help you proceed to planning and implementing your project: Considering my comments above, you'll need to first narrow in on your target audience group. Using your research sources is a good way to do that. Then it's important to (a) be clear on the big picture of your issue, and (b) keep that distinct from possible solutions. It's OK to be looking ahead to solutions, of course. That's our goal with the project. ☺ But as we discussed earlier, concentrating on a single solution could give us "blinders" and keep us from noticing that there are multiple facets to the problem, and likely multiple solutions. You'll need to consider that before deciding what project/action/solution you're going to try to take this semester.

(4) Note sentence-level patterns student should address on their own (5–6).	(4) P.S. Writing technicalities aren't our main focus in this stage of the project. However, if you want to self-assess this draft, here are a couple of thoughts: (5) This draft could benefit from an advanced spelling/grammar checker to catch typos, sentence fragments, missing or extra words, etc. A lot of students find Grammarly (the free version) helpful. (6) Two other ways to catch these types of errors are: a. read the writing out loud, or b. ask someone else to review the draft. These approaches work because a change of perspective, or fresh eyes, can help us notice what we're missing when we've been looking at the draft for so long.
Assessment	This draft was assessed as incomplete/doesn't yet meet expectations. A revised draft with the required components was reassessed as complete/meets expectations.

Line editing doesn't work (at least, not at this point) for two reasons. First, most drafts simply aren't ready for line edits (they aren't the "advanced revision" developmental stage detailed earlier [c2]). Over fifty years of writing studies research demonstrates a central truth: ideas must come before structure, structure before polish (hence the sequencing in tips 2 and 3). Until a writer has a clear sense of what they're trying to say and how they want to say it, grammar, syntax, and the like *must* be incidental. When you're working with most student drafts, the writing is at a stage where it simply doesn't matter if students are using commas versus semicolons or ending their sentences with prepositions. Second, even when drafts *are* ready, line edits foster a view of writing being corrected, rather than writers being developed.

It's not just that line editing doesn't work; as we've suggested above, it's often counterproductive. You've likely experienced this reality. You provide detailed comments and line edits on a developing writer's work. But in the next draft, the only thing that has changed is where the commas are; the underlying issues of flow, clarity, and accuracy remain. It's frustrating—you spent lots of time giving detailed comments, but the writing hasn't improved. What you're experiencing is consistent with extensive research into how people learn to write. A developing writer likely won't connect the line edits you make to the bigger-picture changes needed to move their draft

TABLE 3.2

Example of thematic feedback (research-focused)

Type of project	A brief about the student's research project
Assignment context	Draft proposal, including annotated bibliography. Proposal is developed in the first month of semester, with due dates for individual components. These comments respond to the first full draft (~1,500 words).
Feedback	
(1) Give positive feedback to start.	(1) You have a strong start on your project proposal here. This draft starts to explain why you are interested in this topic. There is room in the next draft to make this even clearer. Good job with your annotated bibliography!
(2) Flag missing components—keep your focus on the draft, not the student. Use phrasing such as "the sources," "the text," etc., rather than "you."	(2) Some aspects of the current draft could use additional work: a. It's not yet clear why farmers will care about native bees as pollinators. Without that, it's not clear why they'll care about the bee sampling method proposed here. b. It would help if the proposal fully explained the importance of the collection methods, in terms of accurately assessing population status and minimizing impact on populations.
(3) Suggest revisions (focusing on only two big things)	(3) In the next draft, it would be good to connect more fully to ideas about what structures pollinator communities, how best to characterize those communities through sampling, etc. Here are some big-picture revisions to focus on first: a. Build out the next draft so it's clear where the proposal is focused on scientific questions and methods and where it delves into the interests of the target audience (you've identified farmers). b. Conceptually, the proposal is coming together. So, your next draft is a good time to focus more on the sentence-level aspects of the text. There are numerous spots in this draft that are wordy or have run-on sentences. It could be hard for a reader to keep track of all the ideas in some of these sentences. To spot these parts of the text, read the writing out loud. That change of perspective can help you notice where you might want to simplify or split up sentences to make them more reader-friendly.
Assessment	This might be assessed, if part of a formal course; or only given feedback, if part of project development with the advisor. In the former context, this draft was assessed as complete/meets expectations (8/10 points). The revised draft, which satisfactorily addressed the clarity, audience focus, and sentence-level issues, was reassessed at 10/10 points.

toward a finished piece of writing. Line edits provide an easy out: the writer can accept each change and see fewer and fewer red marks on the page, but these minor changes don't actually carry the draft into more coherent territory. Worse, it's too easy for a developing writer to accept a change without thinking hard about why it's recommended or what the lessons are for the *rest* of the draft (or for future drafts).

Minimal marking (and related techniques)

So, if you don't line edit, what do you do instead? You need a technique for delivering feedback more quickly while helping the developing writer take responsibility for improving their own work. Fortunately, such techniques exist; and often, from sheer necessity, many mentors have moved a long way toward them already. Think about when you start in on a draft and line edit a couple of paragraphs, then pull back. You realize that you don't have time to continue, and you suspect in any case that the student may not learn much from it. It's a short step from there to line editing just the first occurrence or two of any given error or other writing issue, then instructing the student to "check throughout MS." (It's most effective to give this instruction at the *second* occurrence of the error, as this makes the point that it's repeated.)[19] But one can go further down this road, and we'd like you to try. So, we're introducing you to "minimal marking."

In brief, minimal marking is a practice of making an *X* at the end of each line that contains an error.[20] Multiple marks indicate multiple errors. These marks must be reserved, of course, for categorical errors in spelling, punctuation, grammar, and the like. For text you think warrants revision but isn't explicitly *wrong* (and we have strong takes on that throughout the book), we recommend using other feedback methods.[21] And since positive feedback is essential to a writer's growth, you'll also want to use a ✓ for lines or even phrases you find particularly well written.

In the basic version of minimal marking, this is the only feedback the writer receives for sentence-level issues. The idea is to indicate where writing needs attention without prescribing a fix. Ideally, you couple minimal marking with an as-immediate-as-possible opportunity for students to revise the errors you've flagged. For example, you might return shorter drafts (with minimal marking) with fifteen minutes left in a class session; students then identify and revise the errors before resubmitting at the end of class or at the next class session. In each case, assessment happens only after the students have responded to the minimal marking. Instructors who use minimal marking also recommend tallying the number of marks at the end of the draft and recording them in the gradebook as a way of documenting—for both mentor and student—how self-editing improves over the course of the semester.

Minimal marking coupled with revision before assessment is one of the most efficient tools for writing feedback. It's also a powerful tool for helping students learn to make compositional and rhetorical decisions in their own writing, and to self-edit.[22] As you can imagine, this approach is most effective when other issues (e.g., concepts, flow, etc.) have already been worked out. But if you can't bring yourself to let go of line edits entirely, you may find a happy medium in minimal marking paired with thematic/pattern-level edits.

At this point, you may be thinking, "Wait, really? If the student couldn't address these errors prior to submission, how is this going to help?" Or, as some colleagues have pushed back:

- When peer reviewers do this to us, we absolutely hate it.
- Doesn't minimal marking make succeeding in a writing assignment like pole-vaulting in a darkened gymnasium? Isn't it almost abusive of students?

We recognize these concerns. Indeed, we often hear from folks who, at a minimum, want to use that *X* in combination with "vague" or "watch subject-verb agreement" or "spelling." That way, they give a hint of what they're flagging while still asking the student to apply that hint to the actual text. Isn't this kinder? We hear you. But there are good reasons *not* to sink your time into explaining, even briefly, what the sentence-level errors are. These reasons are at the core of why the disciplines of rhetoric, composition, and writing studies are nearly unanimous about the efficiency and skill-building benefits of minimal marking.

First, published studies suggest at least 60 percent of sentence-level issues are correctly resolved by students after being pointed out through minimal marking.[23] Oftentimes, students are juggling conceptual, organizational, and sentence-level issues simultaneously; such multitasking can lead to syntactical breakdowns, even for experienced writers.[24] Giving students a bit of time to revise, removed from the rush of submitting a text, can help them look at their own work with fresh eyes.

Second, minimal marking avoids the information overload associated with receiving a draft turned entirely red by tracked changes. It also represents a kind of generosity, assuming the student made a mistake rather than that they don't know how to write.

Finally, minimal marking makes it the student's work to correct the error, not yours. This saves you time and cognitive effort, of course, but it also engages students with something a lot closer to what you'd like them to learn: discovering and correcting errors on their own, instead of simply making corrections that someone else has suggested.[25]

Your use of minimal marking needn't end with marks after sentences. You can also gather concrete data on students' patterns of errors and issues, and then design efficient interventions. That may mean some individually focused instruction; for example, perhaps one particular student repeatedly misspells "definitely" as "definately."[26] Alternatively, perhaps many students are showing a common error (like unconventional use of hyphens or apostrophes). You can then point to the shared confusion in the classroom, resolving the matter efficiently.

Don't worry. If you're a fan of editing, or just really particular about certain grammatical or syntactical standards, there's still a place for your close attention in the writing process. Line editing is a valuable contribution to writing, but it needs to happen at much later stages than you may be accustomed to. If we can get you onboard, you'll reserve line editing for coauthoring (or paid editing), and even then, only at late stages of manuscript development.

TIP 5: COACH WRITERS TO DEVELOP SELF-EDITING STRATEGIES.

Writers can learn to identify their own patterns. Fostering this ability is worth considerable investment, as a pattern identified and corrected by the writer is one you don't have to point out a second (or third, or fourth) time. Here are some concrete ways to help writers learn to self-edit.

1. **Advise the writer to read the text aloud.**[27] While many students are reluctant (who are we kidding—many of us are too), you can push them to read their writing aloud as a way to spot unwieldy sentences or leaps of topic. (Lest they be too horrified, you should be clear that you mean reading out loud in private, not in front of you; and as an alternative they can have their word processor read the text aloud to them.) Reading aloud is the most straightforward way to create enough distance for the writer to observe text as if someone else wrote it. It can also help students revise toward a more reader-friendly style in place of the complexity they fall into when attempting to sound "science-y." More experienced students may find similar utility from a change of format, such as reviewing printed text or changing typeface. As part of feedback, the read-aloud suggestion can be motivated by referring to a pattern of errors that could be easily caught this way.
2. **Advise the writer to make an outline for a spoken presentation.**

Many students are more adept at speaking about their work than writing about it. If a draft has major redundancies, gaps in background information, and the like, you can encourage its author to leverage their speaking skills by creating a presentation outline, or even a script. Above all, a presentation mindset helps the student shift their perspective so that they're writing to communicate *with a reader*. This shift is crucial for guiding students to see their writing as an authentic part of disciplinary discourse, not merely as a report back to you of things *you* already know. (See more about the importance of authentic writing tasks in chapter 2.)

3. **Assign reverse outlining.**[28] In early drafts, you're likely to notice that the writer has compressed several distinct ideas into a single paragraph. You may also spot conceptual jumps, with ideas appearing out of logical order or in multiple, scattered locations in the text. While most mentors recognize the value of an outline for wrangling these issues, most developing writers don't—even when they're assigned to create them. Reverse outlines make tangible the value of an outline after a student has created an initial, full draft of a text. They do so by compelling the student to map out what they've actually written versus what they intended to write. You'll ask them to take their draft and reverse outline and identify where the key ideas are, then build a new outline to organize and present those key ideas productively. Then, the student does a "blank-page rewrite," in which they create a new document and pull in only those sentences, phrases, or ideas from the previous draft that adhere to the revised outline. Many students are loath to discard early, weak drafts because of the time they've put into them. Showing them how to build on that work is a way up and out. At the same time, the reverse outline-plus-revision process helps the student move beyond early drafts with a clear sense of what work is needed. This process further facilitates self-sufficiency by helping students discover on their own where weak transitions, redundancies, and content gaps occur in their drafts—and showing them that they *can* address such issues.
4. **Suggest online tools.** Once a draft has progressed to the polishing phase, you can direct the writer to online tools that flag specific elements of sentence structure (e.g., abstract versus concrete nouns, or active versus passive voice). Encourage students not to use these tools prescriptively, though. The point is to help a writer identify their own patterns, and then to make conscious decisions about word choice and sentence structure. There are many such websites and word-processor plugins that you can recommend; we discuss several in chapter 9.
5. **Ask students to develop their own editing tools.** Since the goal here is

to build longer-term writing capacity, not just polish a single text, you can make students consciously responsible for that. Ask students to hand in with their next assignment a personalized proofreading checklist. This is a list of presubmission steps they've chosen based on their previous assignment or draft. Such a list might include, for example, reading out loud, searching their text for words or phrases they overuse or misuse, or using a set of online tools they find productive for their own writing. Even better, you can ask them to justify their choice of the steps on their checklist and provide an example of a change made in response to each. Alternatively, they might provide a "response to reviewers"-style paragraph with their revision. In such a paragraph, students would note the patterns you and they have observed in earlier drafts and how they've addressed them in revision. These responses can even be peer-reviewed by classmates, saving you assessment time and letting students learn from each other.

6. **Assign reflection on the drafting and revision process.** To boil all this advice down: you can help students toward self-sufficiency by encouraging it explicitly. The key is encouraging metacognition about writing. Such metacognitive work helps shift attention—both yours and theirs—to students' growth as writers, and it makes concrete for students what such growth entails on their part. As we keep noting, building a sense of self-efficacy requires, in part, experiencing writing as an endeavor that one *can* get good at. Prompting students to reflect on their own experiences is necessary to improve their awareness of their own improvement. Noting such growth reinforces the time spent to get better; and, more often than students expect, such reflection also enhances their sense of confidence (and even pleasure) about past writing. This shift toward a positive attitude about writing is vital for sustained growth and persistence as a writer.

TIP 6: CHOOSE YOUR FEEDBACK METHOD DELIBERATELY AND TELL STUDENTS WHY YOU'RE USING IT.

There are a lot of ways you might offer feedback on a piece of writing. That's true both of the kind of feedback you're offering (our focus so far) and of the tools you use to offer it.

First: the kind of feedback you offer. If you follow some of the advice above, you might be offering students feedback of a sort they aren't used to receiving. This will work much better if they understand what you're doing, and why.[29] Otherwise, you risk misinterpretation. Perhaps you're using thematic feedback for an early draft and students notice that you made just three comments and two suggestions on an early draft. Without explanation, they may infer—incorrectly—that everything else is perfect. Or perhaps you're experimenting with minimal marking, and students notice that you've done nothing but leave some *X*'s in the margins. Without explanation, they may infer—also incorrectly—that you're refraining from detailed comments because you can't be bothered to take the time to help them learn. You don't need to give them a full-fledged lecture on writing pedagogy, but they'll appreciate knowing that your feedback isn't a secret code, and that it's designed to help them develop as writers.

Second: the tools you use for feedback. Perhaps you prefer to offer feedback directly in a Word file using Track Changes, or on a paper copy with a green ballpoint; perhaps you'd rather write feedback in a separate document coded to the draft's line numbers; or perhaps you'd rather record video feedback (don't knock it until you've tried it). Whatever your system, your students will appreciate your using it consistently and explaining why you do. Not only that, but if you're asking students for specific submission formats (a line-numbered PDF document, for example), transparency about how you're using that format in your feedback will encourage their compliance and, with luck, avoid the impression that you're arbitrarily fussy. One caveat, though: remember that some students may need alternative feedback modalities to accommodate a disability.

Conclusion

While we've focused much of this chapter on feedback that can help a student with a particular draft, you've noticed our larger agenda—one you certainly share. Some of the efficiency in evidence-based writing feedback comes from coaching students into becoming self-sufficient writers who can spot their own recurring patterns, and who can apply tactics gleaned from one writing assignment to others.[30] Acting on our six concrete tips will help you, and the developing writers you support, learn to value the process of building writing skills over the simple "correction" of writing.[31] Making this transition can be disconcerting, especially if you're used to taking a firm hand with submitted text. However, we believe the benefits will outweigh the discomfort of adjusting your feedback habits—because your effort will result in increasingly independent writers who require less of your time and are more capable of applying your feedback.

We also reiterate: through all of this, your attitude is key. It's debilitating to perceive students who submit "subpar" writing as having done so out of disrespect for you or the material. Most students are acutely aware of their poor writing training and would already be better writers if they could get themselves there. Most are also juggling competing demands on their time and must prioritize how much time any given writing task receives. If you understand this, and communicate that you do, you can avoid the kind of adversarial relationship with your students that inflicts unnecessary suffering on both parties. Expanding your instructional methods with the advice in this chapter will let you spend more of your effort on feedback that will make a difference—thus making that feedback more meaningful for you and for the developing writers you support.

4: Writing in the Classroom

What are WAC (Writing across the Curriculum) and WID (Writing in the Disciplines), and why should you care? • What kinds of writing-intensive courses are there, and what are their advantages and disadvantages for helping students write? • How can you design a writing-intensive course that will help students make writing progress without working you to exhaustion? • How can you work toward good writing instruction at the department or program level, not just in your own courses?

Writing takes place in many settings. In this chapter, we're focusing specifically on in-class writing, and we're taking a strong stance: writing instruction should be stitched through science curricula and science courses themselves, not just in exams, term papers, and electives. This chapter makes a case for that integration (including some nerdy historical context), then provides a pragmatic overview of types of writing courses and the assignments that are productive within them. We also address common challenges and ways to mitigate them.

A caveat: the main focus of this chapter is undergraduate writing instruction, because most writing classes are for undergraduates. Graduate students do occasionally have access to writing courses; we encourage this, and instructors of graduate courses can certainly glean useful ideas from this chapter.

Writing across the curriculum and in the disciplines

Before we really dig into classroom models, we'll take a quick detour through the nerdy historical context we promised.[1] Why? We ground many of our recommendations in the research associated with that context; the history can help you better understand why we're making these suggestions. Familiarity with the origins and motivations of two major writing-instruction movements will also help you connect more productively to campus colleagues and resources (see chapter 7).

In the 1960s and '70s, university admission policies in the US and UK became more open. Waves of students entered higher education without elite, preparatory education upon which to base their college-level writing. At the same time, professionals encountered higher expectations for competent writing in the workplace, while lagging literacy levels raised alarms (sounds familiar, right?). The Writing across the Curriculum (WAC) movement was born in response. The WAC idea was to "improve teaching and learning *through writing*"[2] (our emphasis). At the time, this was a radical approach. Rather than traditional lecture-and-exam modes of instruction, WAC embedded writing tasks throughout coursework and posited that students' writing could reflect three things: learning (or discovery), communication, and creativity/self-expression.[3] More contemporary perspectives emphasize that WAC makes writing a tool that instructors can use to help students think critically, learn, and recognize their own role in learning—all while becoming familiar with the expected conventions of writing in their field. This idea is central to how writing-instruction experts talk about writing today.

We're totally onboard with those goals of the WAC framework. Next, we'll discuss why.

WHY THE WAC FRAMEWORK WORKS FOR US (AND HOW IT CAN WORK FOR YOU)

Our case for WAC has a few elements to it. First, a burgeoning body of literature emphasizes what practitioners have said for decades: inter- and transdisciplinary approaches to learning are more engaging and better equip students to address modern challenges.[4] The WAC approach epitomizes how multiple disciplines can be taught together to enhance student learning (versus teaching writing separate from science disciplinary content), especially with respect to writing and critical-thinking skills.[5] Next, we've personally seen writing, when embedded in disciplinary courses, help students learn more and more deeply. Students can then better retain what they've learned and connect it more effectively to later coursework and careers. We've seen students realize this, too. Finally, we've experienced these benefits ourselves: both Bethann and Steve have been trained (or self-trained) in multiple genres and modes of writing. Over our careers, our writing has continually improved as we've learned and practiced the writing conventions of genres as wide-ranging as peer-reviewed publications, popular science writing, journalism, fiction, poetry,[6] and more. Whether or not our own evolution as writers arose from deliberate WAC instruction, students can derive similar benefits from the genre-inclusive WAC approach.

Ideally, though, you'd pair WAC approaches with another major writing-instruction framework: Writing in the Disciplines (WID). In the next section, we'll talk about how we see WAC and WID working together in writing and mentoring.

BUT HOW DO WE GET FROM HERE TO AN ACTUAL CLASSROOM MODEL?

The WAC approach recommends that writing takes place in *all* courses—both stand-alone writing/communication courses and disciplinary ones. It's great if that's already happening where you are, but in our experience it's unusual (and likely beyond your control). If WAC isn't happening in your own department's curriculum, it might be happening in a campus-wide program, such as the Communication across the Curriculum program at Bethann's campus. You might want to tap into such a program, and at minimum, you probably should consider the advice we give in this chapter in the context of such programs. (For just one example of why: some of these programs have specific expectations about what's taught in courses officially affiliated with them.) In any case, you'll want to focus on helping build momentum toward the WAC goal (see further discussion of writing across curricula in chapter 10). In the meantime, there's still plenty of value in integrating writing in a few disciplinary courses, or even in just one. This is likely to be a more focused kind of writing integration, emphasizing help for students in recognizing and deploying the distinct conventions in your field and across genres in your field.

This sharpening of focus—from the broader functions of writing to deploying writing in the typical outputs of your field—is where the Writing in the Disciplines (WID) movement comes in. WAC writing can be expressive or creative and can include a wide range of genres (e.g., learning logs, lab notebooks, or brief reflections[7] or "exit tickets" turned in at the end of class). In contrast, WID writing is generally dedicated to transactional communication involving familiar kinds of writing in science: reports, formal papers, literature reviews, and so on. By "transactional" we mean that the function of these texts is simply to convey information and complete a course requirement; they aren't (usually) creative or a venue for self-discovery or self-expression. These WID assignments represent novice efforts to perform the activity of a professional in your discipline. Both WAC and WID, though, push writing beyond exams, final papers, and other tasks that offer little to no opportunity for iteration and revision based on feedback. Table 4.1 further explores the distinction between WAC and WID in the classroom and explains why you should integrate both approaches.

TABLE 4.1

Our take on how WAC and WID differ, and why those distinctions matter for you and for developing writers (adapted from Townsend 2016)

	Writing across the Curriculum (WAC)	**Writing in the Disciplines (WID)**	**So what?**
Goals	Writing to learn; creativity and self-expression as well as communication.	Writing to communicate.	Both modes are useful in writing instruction; we recommend a mix of assignments to provide students with opportunities for both. Scaffolded writing assignments can start with WAC, then build into WID. (See more on scaffolding in this book's introduction, along with chapters 1 and 10.)
Nature of writing	Writing is exploratory.	Writing delivers content.	Throughout this book, we discuss the importance of providing students with opportunities to explore, draft and then revise. Learning gains are greater when you apply WAC, then WID.
Type of writing tasks	Writing is generally a rehearsal of writing effort or strategy.	Writing is generally a performance of the work of the discipline.	
Role of assessment	Lower stakes; may be unassessed, and/or there are numerous opportunities to revise.	Higher stakes, typically assessed; may not be many opportunities to revise.	
Target reader(s)	Not intended for an audience beyond writer and instructor, though skills gained are intended to be used beyond classroom.	Intended for typical readers in the discipline, though usually the instructor acts as a proxy.	Writing for oneself is meaningful and can solidify learning, illuminate misconceptions, etc. However, WID emphasizes work that's for an *authentic reader*, and our experience and most literature on the subject supports this focus on an actual target audience.
Priorities	Clarity of student self-expression and reflection on learning.	Grammar and conventions of disciplinary writing; content knowledge.	If you start with a fixation on sentence-level "correctness," you risk undermining the work you're doing to facilitate learning. Lean into WAC until students have the content and concepts under control; then teach them to polish their writing (WID).

The bottom line here is that integrating writing in your disciplinary courses is a powerful approach to mentoring developing writers. So, now let's look at how writing can be embedded in your courses, and at the different kinds of courses that can integrate writing.

Advice for any writing-intensive course

We begin with a few points that apply to any course involving substantial writing—whether it's a writing-intensive course in your discipline or a specific writing-focused course.

1. **Embrace the human side of writing.** You and the students you mentor will benefit from acknowledging that many writing issues stem from self-doubt and prior negative experiences with writing and writing feedback. Overcoming these obstacles can involve shifts such as scheduling consistent, regular writing time each week,[8] deconstructing writing projects into tasks short enough to complete in roughly 10–25 minutes,[9] using writing groups for accountability and/or feedback,[10] and divorcing drafting from editing.[11] You can also tap into one of the dominant paradigms in modern writing research, which identifies five concepts integral to improving at writing. As a science-trained mentor, you'll notice these have a specific angle to them that's often traditionally avoided in scientific writing: they acknowledge and affirm the individual human involved. Weaving these concepts into your courses can make writing growth much more accessible and meaningful to students. These "threshold concepts"[12] are: (1) Writing is both a rhetorical activity and a social one. (2) Genres of writing, and the act of writing in them, reflect and influence the thinking of both writers and readers. (3) Because writers have different histories and different writing processes, writing works from experience. The result helps create a writer's identity and can both create and express ideology.[13] (See also box 1.1.) (4) All writers have more to learn. (5) Writing is a cognitive activity. You can leverage these concepts to facilitate student discussion of both negative and positive writing experiences, and then assign writing that initiates writing habits that favor a healthier writing mindset.
2. **Let go of mastery.** Don't worry about polished documents for the majority of the course, particularly if your students are in their first writing-intensive course. Instead, focus on fluency: help students learn to recognize, read, and attempt the range of genres common in your field. You can ask them to draft outlines, excerpts, and other products that help

them recognize and articulate the key attributes of writing in your discipline, but you don't need to require finished, polished texts. This approach dovetails nicely with our recommendation (see chapter 3) that you avoid or greatly reduce line editing. The idea is to alleviate students' anxieties about writing long enough for them to actually learn about writing.

3. **Help students understand that learning to write also involves extensive and omnivorous reading.** Whether students prefer to read thrillers, fantasy, technical manuals about combustion engines, or pop culture-critique essays, every reading experience is an opportunity for them to identify attributes of compelling writing. You can even structure assignments around this practice. For example, you could assign a variety of readings on the same topic: a *Nature* paper, a nature essay, a press release, lab or field notebook entries, poetry, or a short story or novella.[14] Students would then reflect on how different kinds of writing about the same topic are written in distinct ways for different readers. You can then ask students to discuss how they plan to apply those ideas to a future writing assignment with you.

Four types of writing-intensive courses

While the possibilities are endless, four types of writing-intensive courses are common for students in science programs. We introduce them in table 4.2, then discuss considerations for implementing each.

WRITING-INTENSIVE, DISCIPLINE-BASED COURSES (WID COURSES)

WID courses are sometimes known as writing-intensive courses. But "lots of writing" alone doesn't make a WID course, especially if that means weekly lab reports or lots of essay-style response questions on an exam. Rather, WID courses have writing infused throughout the course, with writing at times serving WAC (discovery/learning, creativity/self-expression) and WID (communication) functions (see table 4.1). For example, to make a discipline-based course writing-intensive, you might emphasize disciplinary content while assigning small daily or weekly writing tasks (e.g., questions students still have, one takeaway from class, lab notes), asking students to develop multiple writing products through the semester (e.g., lab reports, outlines, project prospectus, essay responses on exams), *and* including a major writing project as part of the course's final assessment

TABLE 4.2

Four common types of writing courses for developing writers in the sciences

Course type	Pros/cons	Example content	Example assignments
Writing-intensive, discipline-based course	Likely option for most instructors. Takes work to weave writing in without it feeling tacked on or making lots of extra work for both instructors and students.	Content is that of the course: e.g., chemistry, biology	Mini essays, lab reports, exam essays, learning reflections
Writing-focused course inside science programs (usually, "scientific writing")	Most common, writing-specific course in science programs. Time-consuming and challenging to scale for many students. Strong demand from research-focused students.	Professional and scientific writing skills, disciplinary conventions.	Literature review, research brief, grant or project proposal, scientific paper, CV and cover letter
Public-facing scicomm course	Rising in popularity; helps students perceive science degree as transferable skills. Ideally taught by people with expertise in public-facing scicomm.	Public engagement, research impacts in society, rhetoric, psychology of engagement and persuasion, applied science.	Policy one-pager, public engagement-focused project or grant proposal and/or implementation plan, plain language summary, social media plan and content, blog post, K–12 lesson/activity plan, podcast script
Writing and communication course outside sciences, available to science students	Can bolster a science degree for career adaptability without increasing a degree program's teaching load. Least encouraged by many advisers. Science-degree programs may have limited room for "outside" content. Science instructors have limited control over content.	Journalism, creative nonfiction, tech writing, graphic design, public speaking	Science news article or essay, technical report, visualization, speech or presentation

(e.g., annotated bibliography, literature review, draft of manuscript styled for publication, presentation script).[15] All of that might sound familiar—perhaps you or your colleagues already teach courses in which some or all of these writing tasks are assigned. What makes a course an effective WID course, though, is shifting the balance from writing as evidence of content learning alone to writing as a tool for learning both disciplinary material and communication skills. We note that science lab courses can be quite writing-intensive if they have weekly lab reports (or the equivalent) and may even include substantial writing instruction. Such courses are great, but they aren't quite WID courses unless the writing components are varied and more carefully designed for learning as well as communication. (See also our discussion of authentic writing tasks in chapter 2; these are actually rare in lab courses.)

All this expands the scope of your discipline-based course to include the instruction of writing while simultaneously enhancing students' engagement with the disciplinary material. But of course, there are real challenges:

1. **The effort.** This can seem like an insurmountable barrier, but careful design and thought about feedback (see chapter 3) can make the course manageable for you and your students.
2. **Resistance from students.** There may be those who think the writing components are a distraction from memorizing content or narrow career goals. To get these students on board, you'll need to make tangible the transferability of writing skills (see especially chapter 2).
3. **Conflicting instructions from multiple courses.** Students are resentful when what's "correct" in one course is "incorrect" in another—and this resentment can hamper learning. (We tackle this in chapter 10.)

WRITING-FOCUSED COURSES INSIDE SCIENCE PROGRAMS

When most people think of teaching science students to write, they probably think first about scientific-writing courses (e.g., courses focused on "writing up" an undergraduate thesis). These courses can focus on the genres that scientists write in, thereby making explicit the expectations readers have for writing in our disciplines—obvious benefits, right? Ideally, students emerge from these courses able to draft text adhering to the conventions of their discipline, and to polish it into professional-quality text in response to informed feedback.

You might suspect there's a "but." You're right.

Most notions of scientific-writing courses are utopic. People who have taught them often observe that students who are already skilled benefit considerably, average students gain exposure but not necessarily competency, and students who need the most support are further disadvantaged.

Why these less-than-inspiring outcomes? It's at least partly because scientific-writing courses are too often boring—for instructors and for students. We're not picking on our colleagues here. The honest truth is that most instructors we know don't enjoy teaching scientific-writing courses, and we know few students who are enthusiastic about the experience afterward. Indeed, such courses sometimes exist not because faculty are passionate about teaching them but because they feel they *ought to* (perhaps even to stave off the dreaded alternative of the English department teaching "our" students how to write). There's a wasted opportunity here, because teaching students how to write in their field is one of the most powerful and transferable things you can do for them.

Teaching a writing course is hard in part because most of us know good writing when we see it but struggle to explain what makes it so. You and most of your colleagues recognize a competitive cover letter for a job application, a persuasive Introduction or budget justification, or a robustly written Methods section. Likewise, you can recognize when a text is failing at its key tasks. But many instructors and mentors don't have the vocabulary or training to articulate what attributes of the writing render text A more effective than text B. Into this gap, too many writing courses pour an inordinate emphasis on grammar, punctuation, and formatting, and a draconian adherence to dull, formal style that prohibits a writer's voice or flair. They emphasize the polishing side of writing (sentence-level revising) instead of the craft and process of writing—even though the latter is arguably more important and definitely more interesting. It doesn't help that the writing tasks assigned are sometimes abstract rather than authentic. Students might, for instance, write something that's *an example* of a grant proposal but isn't expected to ever *be* one.

So, if you want your scientific-writing course to be more compelling (for you and for your students), you're going to have to do things differently. But how? Here are some straightforward recommendations:

1. Read, dissect, and discuss model texts, choosing examples that represent the range of writing genres in your discipline. Ideally, you'll introduce students to texts that illustrate both commonalities and differences between genres (e.g., lab report, peer-reviewed manuscript, grant proposal, policy brief). Discuss how different genres might treat the same subject for different audiences. Emphasize why (functionally and/or

historically) different genres are distinct in style, structure, and content, and work with students as they experiment with drafting texts that contain the elements readers expect in these genres.

2. Include the genres your students are most likely to produce and interact with after they graduate. Because few undergraduates will ever work in academic settings, a course designed for all majors will misfire if it focuses exclusively on IMRaD papers and grant proposals. (Such a course can be valuable for research-intensive grad students or honors undergrads; as always, you'll want to think about your audience!) A course that serves the broadest variety of students will equip students to recognize and understand the structures of, and draft writing in, multiple genres that are professionally meaningful to the careers they'll pursue (see chapter 2). To make such a leap—particularly if your own background is primarily or exclusively academic—you'll want to engage with folks at the places where graduates are likely to work: nonprofit organizations; consulting firms; industry, local, regional, and federal agencies; K–12 school systems; less-research-focused institutions of higher education; and more. One of the most straightforward ways to understand the range of writing that science-trained professionals do in their jobs is to ask your alumni: *What kinds of writing do you do? Can you share both high- and lower-quality examples of these genres? Would you be willing to talk with students in my class about the essential role of writing in your work?* You can then adjust your instruction to provide students opportunities to learn those common, beyond-academic kinds of writing.
3. Make writing authentic (see chapter 2). Where possible, make assignments engaging by having them be *real*. That can mean any or all of three things. First, writing is authentic if students are writing about their own experiences or about issues they're passionate about. It's worth designing assignments, for example, in which students write op-eds or policy proposals relating to an environmental problem they choose because it impacts the place they live. Second, writing is authentic if the topic or problem it addresses isn't just made up. If students are writing grant proposals, for example, you can ask them to write in response to an actual request for proposals. Third, writing is authentic if it isn't just practice—if there's a real prospect of it reaching readers beyond the course instructor. In Steve's writing course, for example, honors undergraduates and grad students work at writing what will become an actual thesis chapter. Writing of many genres can find public outlets; you can start a course blog, or arrange for op-eds to be published in student newspapers or newsletters of local nonprofits, for example. One note of caution, though: you may

not want (or be allowed) to require that students post their work publicly. Even when this is optional, though, the authenticity is established.

We have fairly specific opinions about how to map out these kinds of learning experiences over a semester, short course, workshop, or in your research lab (see chapter 10). However, including syllabi, rubrics, and assignments would have required a much longer book. Therefore, as a start we suggest you take advantage of resources that can be found in other books or online (see the appendix). In chapter 7, we provide some advice about searching for these kinds of resources more broadly.

COMMUNICATION (SCICOMM) COURSES IN SCIENCE PROGRAMS

While scientific-writing courses emphasize communication with other scientists—or at least folks with substantial scientific backgrounds—scicomm courses emphasize communicating beyond technical/academic settings, including and perhaps especially with the general public.[16] Wherever a student's career takes them, engaging with people both inside and beyond their field will be critical to professional and civic success. That's a paradigm that can lead you to more-impactful and ethical approaches to pedagogy, to more authentic writing assignments, and to students with a sense of science contributing to societal good.

Is this framing utopian, as we noted the idealized scientific-writing course would be? Of course. Increasingly, though, students are coming to us and our colleagues wanting to make a difference to the planet, and society, through science. They can, of course, make this difference—but the efficacy of science in addressing problems is usually contingent upon effective communication. That's true for many reasons. Just to pick two: effective communication helps voters and policymakers see the value of funding science; and the application of science to problems is often, and increasingly, reliant upon collective action by people motivated by effective communication. Ideally, scicomm courses build from a widely held (though not unanimous) interpretation of the social contract scientists have with society—that we as scientists have an obligation to engage with, not just talk at, people beyond the academy. This doesn't mean that every scientist should be leading K–12 school activities or testifying before a legislative body; but it does suggest that sharing science should be part of the training of scientists. All this makes science communication courses a powerful way to gain both administrative support and student enthusiasm (with respect to the latter, see some student reflections in box 4.1).

BOX 4.1

Student reflections on the value of including writing and communications instruction in science curricula

Three students reflect on the value of a science communication course embedded in their disciplinary major:

> "In the past, while I somehow 'knew' what I was supposed to be doing to be more effective, those things may have been an afterthought. I may have approached the communication effort the same way and then perhaps tried to fit in some strategies to make it more effective. Now, I think I am approaching communication through a scicomm lens, so to speak, so that I begin with scicomm principles and try to fit my message into effective strategies rather than the other way around."
>
> *Dr. Amanda Dougherty, postdoc in ecology*

> "Going into this course, I had an extremely narrow view of scicomm as only science writing. I gained skills throughout the semester that I did not intend to, and this is something that has made the class truly invaluable to me. My idea of scicomm is broadened, and I have found that I have a passion for it."
>
> *Quiana Jeffs, PhD student in neuroscience*

> "I specifically mentioned wanting to develop my writing skills in a myriad of ways. My training is mostly in creative writing and extremely formal, submission-type science writing. I'm really glad to say that the different assignments in this class really allowed me to expand my writing into other realms, particularly in the blog-post assignment. I think that writing for other audiences—be that in a blog post or a discussion post for my peers, my project proposal, or in my final poster—forced my mind to sort of expand the way that I think."
>
> *Taylor Wagstaff, undergrad student in wildlife biology*

When people think about teaching scicomm, they may think first about reducing jargon. That matters, but it's not actually where we should start. Sharing science beyond the academy requires instructors and students to consider plain language and vocabulary that's shared, mechanisms for effective knowledge translation including shared values,[17] and skills such as visualization, graphic design, and multimedia work. These things are the foundation of effective and ethical scicomm, which seeks to connect with anyone who wants to use, do, or understand science. It's important for people who do and share science to internalize four key ideas:

1. **The deficit model doesn't work;**[18] that is, people don't appreciate or learn from being told they're wrong or ignorant (see chapter 2). The

problem is most science-trained people have been conditioned to just demonstrate their own expertise and provide information. Scicomm courses ideally help students understand when sharing facts is effective and when being a know-it-all isn't. (For example, throughout this book, we tell you that certain approaches to mentoring may not be as effective as you'd hoped, and we share alternatives that can work better. We're sharing this information in an engaging manner—we hope!—to meet a need you've recognized.)

2. **Science can at best be *part* of social decision-making**, because most people are motivated by many other factors and may feel ambivalent about, or even resent, notions of science as intrinsically important, trustworthy, and objective.[19] Scicomm courses must bring this realization to developing writers who hope the science they do or learn will be useful in society.
3. **Effective and ethical scicomm is calibrated for the specific people** we want to reach and work with. Just as we pitch a paper differently to *Nature* than when we submit to the *Journal of Wildlife Management*, so, too, should we coach developing writers to calibrate their scicomm efforts. Such calibration is especially necessary when developing writers are working to connect with people who differ in background, knowledge base, or values.
4. **Framing science as elite knowledge that should have priority is *not* effective communication.**[20] Any effort to share science beyond the academy must take into account what specific individuals or groups of people know, care about, and can contribute to the discourse and activities of science.[21] That last bit might sound unrelated, but it gets to the idea of co-production: that creating good, usable science means respecting people who are impacted by it and thus should be involved from the beginning.[22] The co-production of science can be a major advantage in setting up science-trained people to share that science effectively.

A great scicomm course leans into these realities by connecting students with conceptual frameworks, literature, and case studies, in combination with opportunities to practice tools that are recognized by researchers and practitioners as ethical and effective.[23] There are many modes of scicomm (e.g., blogs, social media, op-eds, educational outreach, infographics, policy briefs) and possible collaborators and target groups (e.g., community-led science, policymakers, K–12 education, applied medicine). Fitting all those modes and groups into a semester isn't effective pedagogically. And, outside of communication degree programs, even a single scicomm course will probably be a luxury elective, so you'll need to choose a few specific scicomm foci and make clear to students that there's more to the field than your one course.

Clearly, there are a lot of complexities associated with offering scicomm courses. If you're (even partly) responsible for a scicomm course,[24] here are some key recommendations:

1. Root the course in discussion and acknowledgment of the values that all people bring to their interactions with science.[25] In particular, we encourage instructors to engage students with considerations of how science and its communication are affected by matters of identity, credibility, access, funding, colonialism, and more.[26]
2. Take time to discuss with students their concerns and hang-ups around writing and other ways of communicating science. Adding the element of (even potential) communication with the public may change or exacerbate these concerns for some students. Take these concerns into account as you frame the rationale for assignments and the nature of feedback.
3. Identify at most three focal concepts, such as (1) science isn't neutral, (2) you must calibrate to your audience, and (3) because intent doesn't assure efficacy, you must look for evidence of success in communication.[27] Focus all the coursework (readings, writing, project development, assessment) around your focal concepts. If it doesn't build to those, don't do it—at least, not for the first round of the course. A narrow focus can help you keep a course manageable. Narrow focus also helps students work through material in a way that's deep enough to build real capacity with some nuance. Bethann and Steve have both taught survey courses (new project/topic/medium every few weeks), and we find the variety assumes a depth of ability that many students don't actually attain.
4. Make transferable skills the backbone of your course.[28] Perhaps this is obvious; surely we want *all* our courses to be useful beyond the classroom. Leaning into that goal, though, can help your students see not just their scicomm course but their science degree as a bounty of transferable skills. You can achieve this by ensuring that students practice (and revise) multiple types of writing anchored in the same course content. For example, students might outline and draft a report on an issue they want to address, then develop a project proposal and implementation plan, and finally, create a social media post series, policy brief, or other product that's tailored to a specific target group. Doing so models for students how core writing skills underpin the variety of writing tasks they'll encounter as professionals. Meanwhile, students' capacity with writing and with course concepts can be assessed on the basis of how

they account for those concepts in a range of writing assignments. (See box 4.2 for an example of how a scicomm course can be an intensive writing course.)

5. Integrate practice and theory by assigning relevant peer-reviewed literature along with popular and technical examples of effective scicomm. While literature on scicomm (like that on pedagogy!) can be initially unfamiliar to you and your students, it's a key part of ensuring a basis in evidence for what students learn and do.[29] Embed the use of the literature into assignments (e.g., require citations in project development and longer-form writing like blog posts, reflections, and commentaries). Likewise, assign scicomm tools that are used by scicomm practitioners or scicomm-active scholars (such as the COMPASS Message Box[30]). Ideally, you'll also connect students with scholars and practitioners who publish on scicomm. This integrated theory-and-practice model helps students recognize the credibility of knowledge about scicomm while making a scicomm course truly writing-intensive.[31]

BOX 4.2

Example structure for a writing-intensive scicomm course

Bethann has, over time, narrowed in on a specific model for an effective one-semester scicomm course that is quite writing-intensive. The format she recommends is a semester-long project focused on an issue in each student's hometown. It should be a situation the student thinks science could help with but isn't sure how. Their goal in the course is to add something new to the conversation, not just restate the problem.

In-class activities and asynchronous assignments give students practice reading scicomm literature and using common scicomm tools. Meanwhile, students research and write a 5–10-page report scoping out the nuances of their chosen issue. They then develop an implementation plan (including a detailed daily work plan[32]) and do the work they've planned for themselves. In synchronous class sessions, Bethann addresses theory (via readings and discussion) and application (via hands-on workshops on using scicomm tools), thereby modeling integration of these two dimensions of scicomm.

Bethann has taught versions of this course for undergraduates and grad students. These days, she cross-lists the course at both levels to provide a wider exchange of ideas, interactions with career possibilities, and peer mentorship opportunities. She's also taught versions of it both in person and online and has opted to continue teaching it online since fall 2020. (We mention this because it might seem challenging or even impossible to teach this sort of course online, but our experience suggests it can be done.)

WRITING AND COMMUNICATION COURSES OUTSIDE SCIENCE

On almost any campus, many departments outside science offer courses addressing writing and communication. These include courses in English, communications and journalism, cultural studies (gender studies, Indigenous studies, etc.), marketing, and visual arts—as well as those sitting between science and the social sciences (e.g., anthropology, economics, environmental studies, public health). In such courses, students will be exposed to distinct ways of engaging with and creating texts and will learn about the wide range of partners and audiences these fields are attuned to. Courses taught in other disciplines can also offer nuanced engagement with the ethical and humanistic dimensions of science. Such cross-training can make students more adept at recognizing and drafting writing that meets genre expectations in scientific and technical writing in their own field, because they've learned how to recognize and write to conventions in more than one discipline.

Are we being utopic again? You bet. The most obvious downside of such courses is that you can't expect them to be taught by people who are also steeped in the technical and professional contexts, or the conventions and reader expectations for writing, in your corner of the sciences. While skills gained at *any* sort of writing are, in principle, transferable, the transfer doesn't happen automatically. So before recommending a particular course, you might review its content and talk to some students who have taken it (ideally, science alumni who've had enough time to understand how the course did or didn't advance their skill set). And unless you spend some time building a relationship with the department and faculty members that offer the course, you're unlikely to have much influence on the course's content or approach. That reduced control can be difficult for both mentors and science degree programs. To make courses outside your program work, mentors and students alike must be willing to engage with advice, disciplinary norms, and writing styles that differ from those in their home disciplines. Whether you push students to such courses, or they seek them out on their own, you (and your degree program) must find ways to value the cross-training they receive.

There are practical considerations, too. Enrollment caps and course scheduling in other departments may be a hurdle for science students looking to enroll outside their degree program. Furthermore, prerequisites in other departments may prevent science students from parachuting into a single course. Such constraints are reasonable, as students without sufficient disciplinary grounding and writing competence may require much more effort on the instructor's part. They may even disrupt learning for other students while themselves experiencing the course as counterproductive. Science programs must respect the standards and training progressions of partner programs. To do so, science programs may need to allocate additional, sufficient time and credits for science students to take more than one course with the partner program. That will be very difficult in a science department where nonscience minors and diverse electives are nonstarters, but easy where those things are built-in, normalized expectations—as is often the case at universities emphasizing liberal arts-style course diversity. Another possibility is discussion with nonscience departments about what it would take for them to offer a prerequisite-free course for science majors. For example, if a science department provides TA support and a humanities department gains registrants, both departments may benefit. You and your students will get further with a cross-training approach if you engage actively with departments you'd like to send students to. If you obviously respect the values and constraints of partner departments, your students are more likely to get into, and benefit from, their courses.

The bottom line here is that cross-training is valuable because it offers developing writers a multidisciplinary education that we can't fully provide from within science. And pragmatically speaking, directing science students into courses taught by other departments is an efficient strategy. Even though you'll spend time building relationships, you don't have to teach the course (or assess the writing).

Courses, workload, and possibilities

To teach a writing-intensive course or build out partnerships to do so, you'll need to account for the time it takes to design and implement any new course—let alone one that stretches beyond your typical course content. Fortunately, there are ways to reduce this load.

ASSIGNED WORK

In designing any course, it's important to think carefully about what work you're assigning, why you're assigning it, and how you and your students are going to execute it. We say "you and your students" because you all need to fit the necessary work into the rest of your responsibilities—work, other classes, family care, personal time, etc. (see chapter 1 for a reminder of just how much students are likely juggling).[33] This means building in project planning, from daily work plans to high-level consideration of how assignments relate to each other, both for students completing assigned work and for you in designing, coordinating, and assessing it. So that students can benefit from project planning, we recommend you make that an assignment in itself.[34] This can help students grasp the scope of their own ambitions and articulate how they will achieve them.

SYLLABI

It's probably obvious that you'll need a syllabus for your course. Many institutions have requirements for the content of your syllabus; beyond those, there are a few particular items that should be included in a syllabus for a writing-intensive or writing-focused course. These include:

- explicit clarification of the nature of the course and the role of writing in it (writing-intensive disciplinary course versus courses focused on writing)
- candid explanation of why writing is a central aspect of the course, with an acknowledgment that writing is time-intensive (for students and instructors)
- clear and concrete learning objectives for the writing instruction, such that both instructor and students know what to expect and can assess whether the objectives have been met[35]
- identification of any specific resources, texts, or style guides students will be responsible for using proficiently (see chapter 7)

ASSESSMENT AND FEEDBACK

Chapter 3 addresses effective and efficient feedback in considerable detail, so we'll be succinct here. The key points you'll want to keep in mind are:

- **There are decisions to make about expectations and assessment.** Will every student be held to a right/wrong standard for getting an A? Will you take a portfolio approach, with multiple drafts and revisions, evidence of effort, and metacognitive engagement with their own learning all informing a student's grade?[36] Or will you assess each assignment as you might in any other course, with each assignment's assessment contributing to the final grade? How much growth or how polished a project will you expect? Will you assess students in comparison to each other, or in consideration of their development over the semester? We pose these as questions, not as prescriptions, because there are no right answers here.[37]

- **Decide how you'll communicate with your students about plagiarism and about the use of "artificial intelligence" tools, and how you'll deal with these if they occur.** Each of us teaches a range of courses in which plagiarism could crop up. While we don't structure our courses around this, teaching students not to plagiarize (and how and why to avoid it) is important. For further discussion of plagiarism in academic settings, see chapter 8. (It may seem strange and unfortunate that we tackle plagiarism primarily in our chapter on students who are writing in English as an additional language; we'll say more about that decision there.) For advice and caveats about AI tools, see chapter 9, where we discuss them in the context of the wide range of tools that can support writers in practice and learning.
- **Provide detailed rubrics and assignment descriptions.**[38] Rubrics with a fair bit of detail can be beneficial for students as they pursue major writing projects (especially anything longer than, say, three to four pages). They're particularly useful in introductory courses and when your writing course is the only one a student will take in their degree program. Why? Because rubrics help inexperienced students focus more on the content *and quality* of the texts they're writing, rather than just on whether required parts of the text are present. Rubrics make transparent the structures, content, length, and (ideally) any other expectations for a piece of writing. This helps you, too. First, it avoids the enormous frustration of "marking down" students for omissions you feel should have been obvious. Second, it greatly reduces the amount of explaining you need to do when you provide feedback. Instead, you can focus on the nuances of individual students' work while providing feedback efficiently and concisely (see chapter 3).
- **Provide examples of student work, with clarification of how the work was graded.** Succeeding in a course shouldn't be like pole-vaulting in the dark. Students will benefit from real examples of past student work, along with your comments on how that work was assessed and how it does or doesn't meet the rubric. If you're teaching a course for the first time, you may not have such samples yet. You can still use early assignments to provide models for later ones (this is most effective when students have the opportunity to revise or will create the same type of writing later in the course). Be sure to (1) note likely examples as you grade assignments, (2) secure permission from students to use their work in this way, and (3) anonymize the work before showing it to other students.
- **Consider peer review as a mechanism for feedback.** Peer review can be a valuable addition to your feedback process, both saving time and

helping students learn. However, it needs to be actively facilitated so that students understand the goals of peer review and what constitutes useful feedback. We provide a more extensive discussion of peer review in chapter 7.

ADVOCATING FOR WRITING INSTRUCTION

Our advice so far has been focused on you as an instructor in the classroom. But you're also part of, or allied with, a department or other organization. In that role, you can also explore longer-term initiatives that can support writing instruction more generally—perhaps with the involvement of partner departments. Granted, not every instructor will be in a position to make such collaborations happen. If you can at least advocate for them, though, you'll be taking steps toward making (teaching) life easier for you and your colleagues while expanding course options for the developing writers you mentor. Some ideas:

- Advocate for the development of financial mechanisms to offer cross-disciplinary courses or workshops. For example, some universities offer buyouts for instructors from one discipline to cross-teach in a different discipline.
- If your department is lucky enough to have a surplus of teaching assistantships, consider whether they can fund some for qualified students from other departments. For example, English, creative writing, communications, and journalism graduate students might be contracted to help teach writing and communication courses in science programs.[39]
- Argue for your science department prioritizing hiring of faculty and staff with relevant writing, communication, or pedagogy expertise.
- Consider whether your science department, or a set of departments, can develop training initiatives to enhance writing-instruction capacity among faculty, staff, postdocs, grad students, etc.[40]

Parting thoughts

If you've made it through this long chapter, we suspect you share our view that writing instruction in the classroom is both important and challenging. Our advice won't make teaching writing *easy*, but it can make teaching writing *easier*. It can also make it more rewarding—for you and for the developing writers you mentor.

5: Writing Outside the Classroom

How can you guide students to the productive behavior they'll need to tackle the big writing projects typical outside the classroom? • How can you help students understand which features of the scientific literature are useful conventions and which are bad habits to be avoided? • How can you make multiple rounds of revision less painful for you and your students? • How can you help students navigate conflicting advice from multiple mentors? • How can your approach to mentorship change as students gain experience and become your colleagues and coauthors?

In chapter 4, we considered some of the many ways that writing happens in the classroom. For some writing mentors and for some writing students, the classroom may be primarily, or even exclusively, where it's at. But the fact that you're here, reading a chapter called "Writing *Outside* the Classroom," suggests that courses aren't the only writing space where you support developing writers. Indeed, some mentors devote far more hours, over a career, to working with developing writers in nonclassroom settings than in classroom ones—mostly one-on-one, and mostly as those writers draft theses (either graduate or undergraduate).[1]

That's not all of it, of course—perhaps you'll work with students who are writing as part of an independent study or work placement, or with thesis students who are working on non-thesis-writing projects; and you may mentor postdoctoral scholars, junior colleagues, or even peers. Nonetheless, the thesis is the most familiar out-of-classroom situation. It's also one of particular importance, as honors and graduate students are now pursuing bigger writing projects, for higher stakes, than most classes have prepared them for. Writing mentorship can have an enormous impact for these students, partly because of the magnitude of their writing tasks and partly because those tasks are particularly authentic, reflecting their own accomplishments and growth as scientists.

For all these reasons, we focus most of this chapter on the thesis. Toward the end, we also discuss writing with (and mentoring) more senior developing writers like postdocs and junior colleagues. As always, we think you'll be able to extrapolate to related situations. (By the way—as you read, you'll probably find yourself thinking that some of the advice in this chapter would help students with writing assigned *in* a classroom, too. We agree.)

Mentoring writing outside the classroom differs from mentoring in the classroom in many ways, although some are more differences in degree than in kind. Outside the classroom, students tend to tackle writing more independently, without a highly structured syllabus to break big writing projects into assignments with deadlines. Because of the emphasis on the thesis, writing mentorship outside the classroom is also likely to be more focused in terms of genre. There will be more rounds of drafting, commenting, and revision. And you may not be the only writing mentor; there may also be supervisory committee members, co-supervisors, or, later, peer reviewers.[2] Thus, you and the student may have to navigate conflicting advice. And perhaps most dramatic: thesis writing often exists and moves along a continuum from correction to collaboration and coauthorship as the developing writer becomes more independent in writing and in learning (see also chapter 7). We'll address all these differences in turn.

Mentoring the writing process, not just the product

Writing can mean two things (as you'll remember us pointing out in chapter 1). The word "writing" can be a noun (the text that a writer has produced; strictly speaking it's actually a gerund, or noun-functioning form of a verb). But it can also be a verb—the thing one does to produce that text. When you're mentoring a student's writing, we hope both meanings are involved; but the verb sometimes gets forgotten.

If you're like us, you've seen many students who struggle with the writing process, including their behavior. You may even once have been one of them, or perhaps you still are (Steve, for example, battles procrastination every day; at least, he insists that he will tomorrow). Because outside-the-classroom writing typically involves big projects like theses and papers that have extended and ill-defined waypoints and timelines, students often struggle to get started and then to maintain momentum. So, long before you help them settle on content, format, and style, you can help them make a plan of attack.[3] Some suggestions:

- **Talk with your students about writer behavior.** Writing isn't a simple, linear process in which someone taps on the keyboard enough times for a thesis (or paper, or book) to come out. You know that, of course; experience has taught you about procrastination, writer's block, discipline (or its lack), and more. But developing writers may not have formed this perspective yet—and when they struggle, they may feel that they're failing in ways no other writer ever has. So, ask your students what their writing process is like. It may be most productive to do this as a group so that individual students don't feel targeted and can realize they aren't alone in struggling.[4] Do they schedule writing regularly?[5] If so, do they have trouble sticking to the schedule? Do they reward themselves for finishing three hours' work, or a section, or a paragraph?[6] What do they do when they feel stuck? Do they write in private or look for writing buddies or groups?[7] Share your own process and your own struggles. The goal is to normalize both the idea that writers struggle (all do) and the idea that deliberate thought about process is the way through the struggles. You needn't do all this yourself; some writing guides devote considerable attention to writer behavior.[8]

- **Discuss strategies for managing a writing project.** Students may not know where to start, how to build and stick to a writing schedule, or how to dislodge themselves from the writer's block that strikes everyone at some point. A good place to start is, in fact, how to start: students are often intimidated or even paralyzed by the daunting prospect of "writing a thesis." This is normal human psychology, best tackled by breaking the large task down into smaller manageable ones. Discuss with your students strategies for dividing "a thesis" into pieces they can work on with confidence. Don't ask for a draft of their thesis—ask for a rough draft of the Methods, or a paragraph to open the Introduction, or an outline for the Discussion. Notice that we said, "outline *for the Discussion*." An outline is a good place to start, as it has low stakes, can be drafted fairly easily even early in the process, and has the extra benefit of breaking the rest of the work down into smaller chunks that can be tackled later.[9] However, sometimes an outline of the full document is itself a daunting task; an outline for an individual section may be easier for a developing writer to get their head around. Section-specific outlines also provide developing writers with clarity about the distinct functions and conventions of each section of a manuscript. Mapping that out can be productive metacognitive work.
- **Encourage early writing.** Many students think of a thesis as something they'll write after their research is completed, but that misses many opportunities to make the job easier. Writing the Methods, for example, will be quick and easy when they're actually doing the work they're writing about. Parts of an Introduction can be drafted during the literature

review that they'll probably conduct very early on. But it needn't stop there—even a Results section can be drafted before any actual results are in hand, as a way of conceptualizing hypotheses (exploring what results are anticipated and articulating what they might mean).[10]

- **Set deadlines for many drafts and many rounds of revision.** Because "writing a thesis," as a task, doesn't come with an intrinsic structure or internal deadlines, it's easy for a developing writer to find themselves slipping behind the progress they've intended to make. The chunking and early-writing strategies we've mentioned are part of the solution, but even one "chunk" will likely need to go through many rounds of revision. Most developing writers are, at some point, surprised and horrified to find they've sent a mentor three, or five, or seven drafts of a piece of text *and they're still getting it back with more suggestions*. This just doesn't match up with most of their previous writing experience, built on classroom work dominated by writing assignments turned in and rarely revised (and even more rarely revised more than once; see chapter 4).

It's important for students to realize that a major writing project like a thesis won't be done quickly, and that's completely normal; and that they'll be frustrated when it isn't, and that's completely normal, too. We've both seen the load visibly lift from a student's shoulders when they realize their time-consuming writing experience is entirely to be expected. The earlier this realization comes, the more misery can be averted.

From our discussion in chapter 3, you'll know how to structure rounds of drafting and revision for a particular piece of text: start with concepts, framing, and section-scale organization, proceed to organization of and within paragraphs, and move to style, spelling, grammar, and the like only when the big picture is well painted. Talk about a schedule for these iterated drafts, making sure students understand how quickly (or slowly) they can expect your feedback. We understand that you're busy—as we are—but keep in mind that the promptness of your feedback has direct bearing on students' career progression, economic stability, mental health, and more. We know of supervisors who leave students waiting for months for feedback they need to progress; don't be that supervisor.

Finally: the writing plan and associated deadlines are ideally worked out collaboratively, not dictated by the mentor. It's useful to start by having the developing writer propose a timeline, although in our experience those first proposals are usually unrealistically ambitious. Of course, students will miss deadlines, and you'll need to make new ones. You can make future deadlines more helpful by having a candid discussion with the student about why they missed the last one. Reasons might include emotional or mental health issues, avoidance, impossible deadlines, other responsibilities—the list goes on. A good discussion should help you build new deadlines, and perhaps modified workflows, that are more realistic and achievable.

- **Set up an expectation for repeated and substantive revision of what's been written.** Developing writers often tend to view every sentence and draft as precious, struggling to see iteration as a process of thinking. Make it clear that any given draft is likely just part of a foundation—or preparatory work for it—and that it's normal for much of it to be invisible when the final structure is complete.
- **Consider efficient ways of receiving and commenting on drafts.** While we hope you'll take advantage of our advice in chapter 3 to read and comment on multiple drafts efficiently, there's no question that it will take you time to do so. Students are often surprised to learn just how much time their mentors spend reviewing drafts; it's not just their own theses, but their labmates', and manuscripts in peer review, and drafts from colleagues seeking advice. Tell students, then, how to make their drafts easier to comment on. You'll know what you prefer, but a good start is for a manuscript to have indented paragraphs, 1.5 or double line spacing, and both page and line numbers. Do you prefer figures and tables inline or at the end of the manuscript? Would you like a summary of the changes since you last saw the manuscript, plus their rationale (akin to the "response to reviewers" letter that accompanies a revision for a journal)? Do you prefer changes left visible (with Word's Track Changes feature or equivalent), or not? If you're using comments in the margins, should the student remove or leave a comment that's been addressed? All this might sound trivial, but, facing the blizzard of drafts you do as a mentor, making your preferences clear and your workflow easy is anything but.

Encouraging careful thought about the genre and its constraints—both real and imagined

In classroom settings, you may be exploring writing across many different genres. Outside the classroom, a narrower focus on theses and papers can make things much easier, because there are strong conventions to guide writers of scientific papers (with theses usually hewing closely to them[11]) and there are extensive resources to support developing writers in learning them. You may consider choosing a guidebook to serve as the lab style guide, while providing a bookshelf of others for your students to consult. (You'll find an annotated sampling of guidebooks we recommend in the appendix.)

There's an important caution here, though: because the scientific-writing genre is constrained by conventions of content, format, and style, people often feel the need to constrain it even further. In chapter 1 we referred to this as the urge to sound science-y. But our literature needn't be as turgid and tedious as much of it is. There's more latitude in style than either developing writers or their mentors often realize (always respecting, of course, the crucial importance of writing that's clear).[12] With this in mind, imagine that you're reviewing a developing writer's work and something jumps out at you as possibly needing correction or improvement. Ask yourself where that text fits in the following categories:

- Does the text do something that's inconsistent with good writing in general? For example, have you spotted a spelling or grammatical error, or an ineffectively organized paragraph?
- Does the text do something that's perfectly fine in other genres but doesn't meet reader expectations for scientific writing? For example, has a developing writer departed from the IMRaD convention, or used extensive direct quotes from cited material (as is common in the humanities but rare in scientific literature)?
- Does the text do something that's actually very common in our literature but that we'd be better off abandoning? For example, does it use copious acronyms,[13] excessive passive voice, or very long, complex sentences?[14]
- Does the text do something that's common and acceptable in our literature, but you tend not to do it in your own writing? We know you may not want to hear this, but if you're like most academic writers, you have preferences and pet peeves that you're tempted to elevate to rules—but you shouldn't. For example, does the student you're mentoring prefer "crucial" over "critical," while you're the other way around? Or do they include a summary of major results at the end of the Introduction, while you, a murder-mystery fan, hate to give away the ending?[15] We all have preferences; but we sometimes need to recognize that that's all they are.

It should be obvious that writing "errors" in each of these four categories should be handled differently—both in terms of whether you mark them at all (for that last category, just back off), and if you do mark them, how you suggest each be addressed. Careful thought about these categories can also help with a frustrating challenge for developing writers: discovering how to write something effectively while retaining their own voice and interest in the material.[16]

Discussing strategies for dealing with advice from multiple mentors

In classroom settings, a piece of writing will usually receive feedback from a single mentor, and it succeeds if it meets that one mentor's expectations. Yes, there may be conflict between different mentors teaching different courses (see chapter 10), but at least a single piece of writing has to please only one person. Developing writers accustomed to this structure are often discomfited when they discover that their thesis chapters and scientific papers get feedback from multiple mentors: supervisors, co-supervisors, supervisory and examining committee members, external collaborators, peer reviewers, and so on. You know where this is headed, because you've had the experience of getting a manuscript back from peer review and discovering that reviewer 1 and reviewer 2 have made conflicting suggestions. The same thing is almost guaranteed to happen to a student sending thesis drafts to multiple committee members or collaborators. Developing writers—especially students early in their careers—often find these conflicts devastating, believing they're in a hopeless situation in which they can never satisfy both mentors.

Two issues are often intertwined here. First, undergraduate students primarily experience having their work "corrected" and carry this perspective into their later writing; the student assumes there's a single best way to write something, and that their supervisor (or another mentor) is checking to see whether or not they've found it. You'll need to explain, explicitly, that this isn't a helpful way to think about writing. Instead, mentors are collaborating with the student to help them find *a* (not *the*) good way to write something. But that brings us to the second issue: not all mentors understand this either. If you've thought carefully about the kind of comments you make on other people's writing, then you're with us here; but we've all worked with a committee member or peer reviewer who simply expects writing to be done *their* way.[17] If you're the primary mentor (supervisor, most often), you can start by working to reduce the frequency of conflicts among mentors. At least for co-supervisors and committee members, you can—in fact, we would argue you must—have a conversation among mentors to confirm a shared approach to your co-mentorship of writing. Ideally, you'll do this early on, before the student gets caught between differing points of view. A shared copy of this chapter might make that conversation less awkward, or at least help you justify your perspective.

Of course, many writing suggestions will come from peer reviewers or external collaborators who haven't made a preemptive mentoring agreement. And it's entirely reasonable for two mentors to disagree about a point of writing, and for each to state their case. So, solutions need to involve the developing writer, too. When your student is puzzled by another mentor's comment, you can help them assign that comment to one of the four categories above (see "Encourage careful thought about the genre and its constraints") and thereby help them decide what to do about it. Alternatively, you can suggest they ask the mentor who made the comment to clarify how they would categorize it. Perhaps most important: you can explain to the student that ultimately, it's *their* thesis, *their* paper, and they can choose not to accept one mentor's or reviewer's suggestion—if they can explain their rationale for that choice. They may be surprised to learn that in a way, conflicting suggestions put the writer in a position of power. If mentors Ana and Bo want opposing things, then the writer can take Ana's suggestion, or take Bo's; or, perhaps best, do something different altogether. Explain to your student that in this situation they can judge for themselves which course of action is best—and they can then explain their choice to Ana and Bo. A student's realization that they can and must make such judgments is a normal (and vital) part of their growth from being "corrected" to being a scientific collaborator.

Once a student realizes they don't have to execute both of a pair of conflicting suggestions, it's a short step to realizing they don't have to take every suggestion even in the absence of conflict. Of course, the key is still that they must explain cogently why they aren't following a mentor's suggestion. Keep in mind that this applies to your own suggestions, too. A good mentor is happy when their student first pushes back against a suggestion, and even happier when their student first wins the argument.

We've focused so far on mitigating the problems that arise from multiple mentorship. But while too many cooks sometimes spoil the broth, many hands can also make light work. The difference is a plan. When a thesis (or chapter, or paper draft) will pass through multiple supervisors and/or committee members, consider dividing the work. Two supervisors may take turns dealing with the earliest drafts; or one may be more suited to reviewing content and another to reviewing structure and style. What everyone should avoid is the situation where multiple mentors waste time making the same comments on the same rough patches. When a single draft goes simultaneously to multiple mentors (as is common in later stages with supervisory or examining committees), consider making a plan for division of labor; perhaps not everyone needs to look for comma splices. If you work repeatedly with the same co-mentors, you'll learn over time how to best take advantage of complementary interests or expertise, and how to manage everyone's pet peeves and habits.

Working with more senior developing writers, and the transition to full coauthorship

The relationship between mentor and student changes over the student's development as a writer. Eventually, developing writers should gain the confidence and ability to write, and to continue their development, on their own. This transition is likely to be gradual, but when it's complete, you've seen a writer shift from a student into a colleague and coauthor.[18] This should be a source of joy and pride for both parties.

You might wonder how, as a mentor, you can help this transition to independence happen faster. Tuck that question away—we'll tackle it in chapters 6 and 10. For now we'll assume that, for at least one developing writer you deal with, the transition is underway, and we'll suggest ways you can provide them with appropriate and effective support. Perhaps you're working with a senior PhD student, a postdoc or research associate, or even a colleague who's recently joined the professorial ranks. You won't be surprised by where we start here: the most important thing is a clear and shared understanding of just what the writing relationship is. At what stage of development is the writer? What kind of participation in writing, from the mentor, is appropriate to that stage? Your key to efficient and low-stress collaboration is to make sure you and the developing writer agree on the answers.

We've beaten the drum pretty hard about moving developing writers away from being corrected and toward doing the metacognitive work of recognizing their own writing limitations and looking actively for solutions. That's an important motivation behind minimal marking (see chapter 3), for example: to shift responsibility for finding solutions to writing problems from mentor to student. It's also why we urge you to suggest changes rather than make them or require them. But with writers further along in their development, you can gradually relax this attitude. At some stage, it's no longer necessary, or efficient, to mark a passage as needing revision instead of simply revising it. It's also no longer respectful or appropriate to comment or revise text as if you're the guru and the arbiter of "correct" writing. Does that sound contradictory? Not if you see it as recognizing the increasing confidence of the advanced writer, and their understanding that when you edit, you're not correcting—you're suggesting. Moving toward the kind of editing you'd do with a coauthor is recognizing the advanced writer's ability to find mentoring even in edits. Of course, except with the most advanced writers, you'll still want them to notice those edits. Tools like Word's Track Changes are perfect for this, although you may need to emphasize that you'd like the developing writer to consider (accepting or rejecting) each edit, rather than clicking the oh-so-tempting "accept all changes" option.

The changing practices we're suggesting here will work best if you and the developing writer both understand what you're doing, and why. You might, for example, explicitly tell your nearly finished PhD student that you've come to think of them as a collaborator as much as a student, and therefore you'll take a different approach to writing than you did earlier in their development. You might say "Because you're now a much more adept writer, I'm going to start just making edits—but please think about why I might have made each one and consider whether you want to keep it or make the case for something else." This approach seeks to build on the developing writer's maturity to achieve further mentoring with greatly reduced effort on your part.

Eventually, you'll realize that it isn't just that they're a collaborator *as much as* a student—they're a collaborator *instead of* a student. This is wonderful evidence that your mentorship (and always, the mentorship of others as well) has been successful. Someone you once thought of as a developing writer, and in whom you invested lots of mentoring effort, has become a colleague with whom you coauthor manuscripts and other documents as equals. Now you'll be working primarily to improve the current manuscript rather than to mentor the person writing it. You might breathe a sigh of relief at this point. It's a lot easier, after all, to polish a piece of writing than it is to mentor a writer. But we encourage you to remain intentional about how you work together. None of us is ever done mastering the craft of writing; you can now celebrate coming to the point where you learn as much from your new colleague as they learn from you.

Managing writing projects with colleagues is beyond the scope of this book. Nonetheless, we offer a few suggestions for working with the early-career writers we're talking about in this section. An early-career writer may have less experience as part of a co-writing team, and so careful discussion of the division of responsibilities and about timelines for completion will pay off. But you might be pleasantly surprised to find that the early-career writer is now driving this discussion. In fact, co-writing projects benefit from having someone in a role we can call "lead writer," and the early-career writer may be the best choice for this. The lead writer is someone who oversees and steers the project, keeping track of each co-writer's contributions, maintaining a definitive version of the manuscript as multiple authors make changes, editing to maintain consistent style and voice, and handling submission for publication and subsequent communication with the journal or other publication venue. This role is well suited for someone good at tracking details in a large project and keeping to a schedule.

Summing up

Mentoring writing outside the classroom can involve mentoring so intense, and of such long duration—think of the PhD thesis—that it can feel a bit like parenting. Like parent and child, mentor and student may be heartily tired of each other at times, and yet bonded for life in appreciation (or trauma). We think the advice we've offered here can sway the balance toward less tiredness and more appreciation. In fact, while we've addressed our writing to the mentor throughout this book, this is one chapter we strongly suggest be read by the student, too.

There are few feelings better, for a mentor, than seeing a student flourish on their own. No matter what their career, the students you mentor in writing will build on your mentorship in many exciting ways. That's why we all work so hard at it—and why we celebrate when our students leave the nest.

6: Teaching and Mentoring Toward Independent Learning

What does it mean for a writer to begin learning on their own? • How can you help your students reduce their reliance on you (and on other mentors)?

You probably have an interesting dual perspective on the students you mentor. Seen from one angle, they're always the same: last year's class enrollment or cohort of new students is replaced by this year's, and this year's will soon be replaced by next year's. The mentoring can seem the same, too: you've just managed to convince one cohort that scientific writing can (and should) use the active voice when another cohort arrives and needs to be convinced of the same thing. But seen from another, less frustrating, angle, individual students move through their program and develop as increasingly skillful and confident writers. It's easy to let the former perspective overshadow the latter, but the latter is more important—and it's a window on a longer-term process of development that's worth thinking about.

Here we'll take a deep dive into that process, focusing on writers at a quite advanced stage. Graduate students well into their degrees are the most familiar example, but you can apply a lot of the thinking here to writers at other stages. In fact, you'll notice strong echoes of our advice in chapters 3–5, where we recommend mentoring practices that encourage students to take more responsibility, and do more thinking, about their own development.

The long path of a developing writer—and why it matters

If you could watch an individual writer go through their entire development, you'd see a long process with many different mentors using many different instructional and mentoring techniques. Perhaps it began with an adult helping a toddler hold a crayon to trace out their name. Later, grade school teachers assigned fill-in-the-blank worksheets and short-answer questions. In high school, short answers were extended to essay responses and longer-form projects; in college, lab reports and term papers joined the mix. That wasn't the end—perhaps there was graduate school, with its usual emphasis on writing scientific papers and the thesis, or perhaps another career (see chapter 2), with the developing writer increasingly expected to write independently, to lead a writing team, or even to oversee and mentor more junior writers. Critically, this isn't just growth in *writing*; it's a process of becoming adept at using writing in *learning*, *thinking*, and *communication*. Early on, a student needs continuous, detailed instruction and is likely to conceptualize writing as a task that produces a particular product (a filled-in blank, an answered question, a term paper). Later, if all goes well, a student transitions to seeing writing as a craft they're practicing and at which they can continue to improve, eventually becoming both self-motivated and self-mentoring. This—independent learning by the developing writer—is the goal.

Why sketch out this process here? Because it helps establish three important things about the way we teach and mentor toward independent writing.

First, it's far more important to help a developing writer find the tools for their own independent learning than it is to teach them any particular thing about grammar, format, or style. After all, we simply can't get to mentoring the next cohort of developing writers unless we can equip the current cohort to write, and to develop as writers, without needing our continued advice and intervention.

Second, different mentoring approaches are needed at different stages of a writer's development. This is obvious if you compare writing help for kindergarten students and PhD candidates. It might be less so if your mentoring focuses on a narrower slice—say, junior and senior grad students, or first- and second-year undergraduates. It's still true, though. (In chapter 10, we explore some ways that writing instruction might be scaffolded through a curriculum or lab group to reflect the changing needs and abilities of writers as they develop.)

Third, any developing writer you see has had a long prior history with other mentors. That's essential; none of us, alone, has the tools or the time to bring each new writer from square one to complete independence. But it's also a potential source of complications, because you'll often deal with writers who have received well-meaning but misguided advice from previous mentors. This is perhaps most painfully true with respect to scientific writing, where most of us have colleagues who still teach dated practices like exclusive use of the passive voice, or who reward students for mimicking the worst attributes of the existing literature (e.g., stiffly formal phrasing, deluges of acronyms, complex sentences, and the longest possible words).[1] As an evidence-based writing mentor, you'll need to be prepared to advise students to reject some of what they've previously learned. This can be tricky, as they can resist and then understandably resent finding out that they've been misled. You'll need to explain why your advice should be taken over the earlier advice it contradicts. (In chapter 5, we discussed resolving conflict among mentors working at the same time; the issue of overcoming previous mentorship is clearly related, and it's key to both that students realize they need to weigh advice before taking it. In chapter 10, we return to the mentor-conflict issue in the specific context of designing writing instruction through a curriculum.)

What do we wish developing writers would do?

If you're like us, you've occasionally felt frustrated as you make the same writing suggestion (Use the active voice! Think about your topic sentences!) on draft after draft of a student's work, and then again for student after student. You plead with any deity that might be listening: couldn't students learn some of these things on their own? They can, of course, and that's really the point of the long developmental process we sketched in the last section. Your goal, and ours, is to bring students to the point that they no longer need us—where they write and, beyond that, learn independently.

But first, what do we mean by writers who learn independently? We don't mean writers who never learn from anyone else. We doubt that *any* writer, no matter how accomplished, goes it alone; all of us need peer review and editing, and all of us keep learning, no matter how proficient we get. Instead, we mean writers who actively seek help from many places (not just from a single mentor), and who deliberately take advantage of this help to advance their craftsmanship in ways they've decided to target. The transition to independent learning won't happen all at once (and we remind you again that much of the advice we offer in chapters 3–5 is shaped around early steps in the transition, or at least laying the groundwork for it). So, if you're mentoring a writer who doesn't leap immediately from needing constant intervention to handing you mature, polished drafts without your help, don't panic, and don't give up. While some birds fledge quite suddenly, writers can and do take wing more gradually.

The skills and practices used by independent, self-mentoring writers aren't easily reduced to a list. Nonetheless, we can suggest a few key things that mentors wish developing writers would do—or to phrase that more constructively, a few key things that mentors can help developing writers begin to do. An independent learner can learn as well as generalize from advice. If you return their draft with the comment that the content of the paragraphs ought to match their topic sentences, their revision will fix that not just in the instance you circled but throughout—and their next writing project will arrive as a draft without the same problem. If they don't understand your advice (perhaps they don't know what a topic sentence is), they'll research the issue rather than just asking you about it. An independent learner takes advantage of resources like writing books, blogs, writing centers, and webinars—at first, perhaps, those you point them to (see our recommendations in the appendix), but later, ones they've found themselves. They talk about writing with their peers, comparing notes about challenges and how they've overcome them; they may even set up writing groups to formalize these discussions.[2] They read broadly and reflectively, looking for characteristics of good writing to model and characteristics of bad writing to avoid. And although perhaps this isn't strictly a component of self-mentoring, they begin to pay it forward, mentoring the developing writers around them. They may, for example, offer advice about writing behavior or comments on drafts to their more junior labmates, thus displacing some of the work that would otherwise have fallen to you as the mentor of them both.

There's more, of course—the job of a writer who self-mentors as they practice their craft is a multifaceted one. You can identify elements of self-mentoring as you execute them yourself, but it's much more likely you'll identify them when the developing writers you mentor *don't*. It can be tempting to resent giving support you don't think you should have to provide (and to commiserate with your colleagues who are nursing their own, similar issues).[3] But resentment won't help. Instead, keep in mind that transitioning to independent learning is a big job, and everyone needs help before they can fly. We did, and you did, too.

Can we arrange for our own obsolescence?

It's perhaps tautological: when a developing writer becomes an independent, self-mentoring learner, then they no longer need you. In other words, your goal as a mentor is to sow the seeds of your own obsolescence.

Before this, though, if students aren't learning on their own, it shouldn't surprise us. There are three important obstacles between most students and where we'd like them to go. First, they don't conceptualize that they should be learning on their own. They remember a long history of having their writing "corrected," and so that seems normal to them; and they imagine gradually getting better through correction until they don't need to be corrected as often.[4] One of the major benefits of moving away from line editing of student writing (as you'll remember from chapter 3) is that it can help students move past this passive approach to learning. Second, even if students are resolved to learn on their own, they don't know what they should learn or how to learn it. There are many resources available to them—to diagnose their own challenges and find solutions to them—but few students seem to discover them. And those who do often think they're too busy writing to spend time with resources that might help them in the longer term. Finally, we as writing mentors often don't know how to tell our students what and how to learn, because none of this was made explicit in our own training, either. Your realization of this gap is probably a big part of why you're reading this book.

Those are steep obstacles, but they aren't insurmountable. The path forward lies in mentor and student deciding together that the goal is for the student to become an independent writer—and then talking about what that looks like and how to get there. What can a student do independently now, what should they be able to do, and how can they make the progress that's needed? To this end, the mentor must have clarity about the work they're asking students to do, and the vocabulary to express it: learning content and writing skills, learning to use feedback, learning to take the lead on a writing project, and so on.[5]

A program for getting there

You can make an enormous difference with a relatively simple approach. The key is in three parts: (1) motivate the developing writer to want to take responsibility for their own learning; (2) identify the skills and practices you'd like the developing writer to adopt that will free them (gradually) from your support, and make them aware of them; and finally, (3) help the developing writer to make and follow a plan for that adoption. But how do you do these three things?

Let's take motivation first. A student might suspect that you're nudging them toward independence simply to offload the hard work of helping them write. *We* know that's not what you're doing; but you should make sure they do, too. Emphasize that almost no matter what career path they take, writing will be a major part of it (see chapter 2). Tell them explicitly that your goal is to make them a better writer forever, rather than just improving the writing they're doing right now.

Now, moving on to skills and practices: we're going to suggest something that might sound condescending or awkward—but it needn't be either of those things if approached with empathy. Table 6.1 provides a checklist of

TABLE 6.1

A checklist of skills and practices for a developing writer working toward independent learning

Skill or practice[a]	Status (yes, no, maybe, don't understand)
1a. Name two challenges you face in your own writing behavior.	
1b. Suggest a solution to each behavioral challenge.	
1c. Implement each solution.	
2a. Develop a concrete writing plan (intermediate self-determined deadlines, daily writing schedule, etc.) for a writing project.	
2b. Follow through with the plan in 2a until the project is successfully concluded.	
3a. Name two support people (librarians, etc.) you could ask for writing help. Identifying a role ("librarian") is OK; identifying a specific person ("Jalene, the reference librarian at the engineering library") is better.	
3b. Identify two specific tasks a support person could help you with.	
3c. Meet with a support person for writing help.	
4a. Identify two books of writing advice suitable for your discipline and stage of development as a writer.	
4b. Read those two books and make notes on at least three practices you will adopt or change as a result.	
4c. Having implemented the three practices in 4b, write a reflection on their effectiveness.	
5a. Identify at least one writing authority/expert (in your discipline or not) who has a podcast, website, blog, or social media presence.	
5b. List at least three suggestions for your writing gleaned from the authority in 5a.	
5c. Having implemented the suggestions in 5b, write a reflection on their effectiveness.	
6a. Receive comments on a draft, and make a checklist of errors/issues to avoid or self-revise in future writing.	

6b. Revise a draft using the checklist from 6a.	
7a. Read a paper or other piece of writing and make notes of two things to emulate and two to avoid.	
7b. Act on the plans from 7a in your own writing.	
8a. Form a writing group to discuss writing practices, review/edit each other's writing, or read critically together.	
8b. Meet with a writing group at least six times and write a reflection on outcomes from this practice.	
9. Write a draft and then make substantial revisions (not just edits) before submitting to writing mentor.	
10a. Identify a lab mate or other close peer who could give useful comments on a draft.	
10b. Receive comments from your peer in 10a and revise as a result.	
11a. Identify a more distant colleague (peer or more senior) who could give useful comments on a draft.	
11b. Receive comments from your peer in 11a and revise as a result.	
12. Add two more skills or practices to this checklist that would advance your independence.	
13. Reflect on a time when you've enjoyed writing. When was that, what kind of writing was it, and how can you build some of those factors into the writing you're getting better at now?	

For each row, under "Status" the writer should indicate "yes," "no," "maybe," or "don't understand" before discussion with the mentor. The checklist exercise should then be repeated every 6 to 12 months to measure and celebrate progress. An editable version of this checklist is available at https://www.helpingstudentswrite.com/resources.

a. Blank rows are included (and can be filled in and multiplied as needed) to remind the mentor that this list is far from exhaustive. Skills and practices are described somewhat vaguely; this is deliberate because there is value in the developing writer thinking about how to interpret them.

skills and practices that represent steps along the transition to independent writing and learning. It's not exhaustive, of course; no such list could be. However, we've included skills and practices that, in our experience, are either powerful steps forward or steps students are often reluctant to take. Our suggestion is to give a copy of this checklist to a student who's early in the transition to independence. Ask them to work through the checklist, with each box receiving a "yes," "no," "maybe" or "don't understand." Then—and this is key—you and the developing writer will sit down and work through the checklist together, talking about the skills and practices it represents and making a concrete plan for a future repeat of the exercise to involve more yeses and fewer noes. Many students will need to be pushed a little to be very concrete about the checklist items: if they report having taken an action, ask them to describe specifically why they chose it, what they did, how they think it worked out, and what next steps they will take as a result. The more active and detailed they are as they think about, plan, and pursue their own development, the better.

This exercise can accomplish several things at once. It helps you (forces you) to be deliberate and explicit about what steps you're asking a student to take. It can make the student aware of particular skills they lack or practices they ought to implement. At a more metacognitive level, it can help the student realize that there *are* such skills and practices, and that they can take stock of the contents of their toolbox. It can also give them a means to fill that toolbox in concrete and trackable ways. It can reduce a mentor's frustration, because you, too, can see progress. Finally, it can build confidence by showing the student that they aren't starting from square one. Nearly every student will fill in a few yeses; if you customize the checklist, we encourage you *not* to remove capabilities you think your students already have. (You're aiming here to help students to identify growth and "wins," not just develop a herculean to-do list.) The guided conversation that results from repeating this exercise over your time with a developing writer can be powerful indeed.

Wait, it's not all on you, is it?

Definitely not! We've focused on your one-on-one mentoring of relatively advanced writers here, because in our experience that's when a lot of rapid growth happens. We remind you again that a lot of the advice in chapters 3–5 revolves around setting the stage for this growth. For most developing writers, the transition to independence won't come all at once and will involve multiple mentors. Consider, for example, a reflection by one of

Steve's PhD students, Rylee Isitt (box 6.1). Rylee points to several different undergraduate experiences that brought him from rote learning and performing homework up through independence as a scientific writer. (You'll notice also that Rylee emphasizes writing experiences being "real"—a major theme in chapter 4).

BOX 6.1

A PhD student's reflection on writing independence

In my undergrad, scientific writing was a major aspect of most second- to fourth-year classes. These were effective at teaching what sections journal articles have, what goes into each section, how to cite sources and statistics, and so on. But no matter how seriously I took it, I was aware that it was "homework," not a real job. Most of this work was completely dictated to us: we did the procedure in the lab manual and wrote a report following a detailed style guide given to us by the instructor. These courses taught writing by rote and didn't prepare me to think independently.

One course was different. It was an upper-level animal behavior course where we had to design a real experiment, get animal-care approval, go into the field to collect real data, and run analyses to test our hypotheses. We then put together a poster for a mock conference. It was exhausting, but it felt quite "real" even though we knew the chance of publication was slim. The experience prepared us to think (and write) independently. Around the same time, I assisted a PhD student as a lab technician and helped her write parts of her journal article. This, too, was extremely helpful. It gave me much-needed experience with the real scientific and peer-review process.

When I started my MSc, despite my limited experience I felt like I'd already done some "real" writing, and I wasn't intimidated by the sudden independence that was given to me. I'd experienced it before.

Nothing beats learning by doing the real thing. But even undergrad courses can get very close to "real," giving students the experiences that they need to become confident writers.

Rylee Isitt, University of New Brunswick

Endings, and new beginnings

When all this works, you'll find that the developing writers you mentor need less and less of your attention and your time (thus making room for the next cohort to claim your attention). Make clear to these increasingly independent writers that you're not abandoning them; instead, you're recognizing and celebrating their growing independence. In addition, their time under your wing may be coming to an end, but that ending is also a new beginning. Often a developing writer who used to be a student is now a collaborator. But even when someone you've mentored lands further afield, you'll know that they're now equipped to continue their growth in the craft through their career, and to mentor other developing writers in their turn. Take a moment—or many of them—to feel the pride you deserve for helping them get there.

7: Sharing the Workload (and the Fun)

How can you pull in resources to lighten your load in mentoring writing? • There are dozens of books on writing available—which might your students benefit from reading? • How can you find effective writing exercises to use with your students? • How can you recruit colleagues to help you mentor students in writing? What about campus resources such as writing centers and libraries? • How can you use peer review and other activities to help students learn and grow together?

It's not a fantasy that mentoring developing writers—either in the classroom, or beyond it—can be meaningful. Sometimes even, dare we say, fun! But for many reasons, you might still shrink from the prospect. We dug into these reasons in detail in chapter 4; in brief, they cluster into issues of time, expertise, confidence, and interest. Fortunately, all four of those issues can be addressed with one key realization: as a mentor, you don't have to provide writing support all by yourself. It can be much easier to provide what developing writers need when you tap into resources and strategies that reduce your workload and simultaneously enhance student support.

Does that sound like a big promise? Read on!

The resources you need are out there, and more easily available than our colleagues often realize. In brief, you can lean on (at least) the following: written resources such as syllabi, writing guidebooks, and exercises shared by others; your peers and colleagues; writing centers, librarians and the library resources they steward; and your students' peers (and near peers).

Written resources

We'll start with a straightforward option: written resources, including books, writing exercises, and syllabi. Each type is plentiful to the point of being overwhelming. Which of the many options should you use, and how can you decide?

We'll walk you through how we make these decisions ourselves so you'll have a heuristic you can use to make decisions. But to make things simpler, we also provide some direct recommendations.

WRITING GUIDEBOOKS/SELF-HELP BOOKS

Publishers' catalogs, library shelves, and the internet are bursting with writing self-help books—some aimed at all writers, others more narrowly tailored to scientific writers. Sometimes these books are distributed to new faculty during university orientations or at training events, and scientific-writing books are often featured in exhibit halls at professional conferences. When you yourself were a developing writer, perhaps you turned to a writing book—or at least you felt you ought to, so you bought one. But if you don't own one, or never read the one you have, you're not unusual. Many of our colleagues, and many of our grad students, are convinced they're too busy actually writing to read a book about writing. But here's the thing: building strong writing skills (through practice, feedback, and reading) is one of those important investments that pays back throughout the rest of your life—almost no matter what career you find yourself in (see chapter 2). So, we urge you to take the time to dig into a book or two of advice; we hope you'll urge the developing writers you mentor to do the same.

When the writers you mentor spend time with a writing book, it can pay off in at least two ways. First and most obviously, good books really do help—and everything a book teaches a developing writer is something you don't need to teach them. Second, you can use the advice in a writing book to simplify your feedback and to standardize and communicate your own expectations (see chapters 3 and 10). When you and your students settle on a single book as an authority on best practices and matters of style, you reduce both your cognitive and time loads substantially. You also help students gain confidence by reducing their concerns about whether they're writing something you "like"—because you've demystified that right up front.

Which book, though? There are many to choose from; some are more helpful than others, and which ones are most helpful may vary across writing projects and career stages. We know you don't have time to read and evaluate all the possibilities! So, we provide some specific recommendations below and more in the appendix. However, if you'd rather not take our word for it, here are some criteria to help you identify a book to recommend to the developing writers you mentor.

1. Shorter books may be the most useful to early-stage developing writers.

Students are more likely to use a reference that's brief and digestible, especially if it's written engagingly. Shorter books also tend to be calibrated to writers looking for an accessible combination of advice and exercises to reinforce it. Finally, smaller books are more affordable, and it's worth lowering the barrier to the resources you require or recommend (see the introduction and chapter 1).

2. Narrow your search to books that cover the genres you're dealing with.

You can save yourself and your students time by checking the table of contents of prospective books. Does the book cover the kinds of writing products you're working on with your students? If not, keep looking. If you're choosing a book for a single course, don't make your students buy a massive, comprehensive reference only to use a few of its chapters (but see chapter 10 on the possibility of adopting a single book for use across multiple courses).

We've harped on writing being a transferable skill (see especially chapter 2), and of course writing advice can generalize across tasks and even genres. But developing writers, particularly at the undergraduate level, often haven't yet built the skill set they need to understand advice as transferable. Our advice about choosing an on-topic book reflects this developmental reality. Steve's writing-advice book, for example, addresses writing academic papers in considerable detail, but grant proposals get a scant three pages.[1] While Steve assumes that many readers of his book will be able to extrapolate advice on papers to enhance their grant writing, asking early-stage developing writers to extrapolate in this way may push them into the deep end before they are competent swimmers. The experience is likely to be frustrating and the indirect advice underused, plus you risk reinforcing students' worst perceptions of writing as mysterious and unrewarding. Thus if your main goal is to teach grant writing, use a book about grant writing. For another example: most scientific writing-advice books generalize quite well across different STEM fields, but writing in mathematics (especially pure math) is quite different, suggesting the use of a math-focused book.[2]

3. Lean on recent books and newer editions.

Your students need to be up to date in writing as much as in their disciplines. Writing practices do evolve, and contemporary expectations may differ from older ones—for instance, in terms of science ethics, norms for specifying Methods, and even linguistic, grammatical, and stylistic patterns. Recently published books can help you identify (and let go of!) your own out-of-date preferences. We all have them. We offer one counterpoint, though: *very* recent books are less likely to be available used, and that can make them less affordable.

While following trends isn't always desirable, a book in a second (or further) edition sends a useful signal: it's been received well enough by the community for the author and publisher to invest in an update. As a bonus, used copies of earlier editions may be especially affordable. You'll want to consider what's changed in the newer editions, of course; sometimes the changes are significant. Indeed, comparing advice between editions is a valuable strategy for you and your students. What changed? How does the author explain these changes? If they haven't, can you discern a reason for the updates, perhaps in the way disciplinary norms and standards are evolving?

A few recommended books

For a short, student-friendly book especially useful in the undergraduate classroom, we recommend Anne Greene's *Writing Science in Plain English* (2025). This book is succinct, affordable, and accurate, and provides useful examples across a range of science disciplines. However, it's not a detailed guidebook to scientific writing as a genre, and it doesn't engage much with the psychology of writer behavior. If those are important kinds of coverage for you, then we recommend Steve's *The Scientist's Guide to Writing* (2022). Each book also provides dozens of writing exercises that you can assign directly or use as reliable models for developing your own.

We should also mention that ours is not the only book about *teaching* writing. Perhaps it's not quite apt to describe writing-instruction books as sharing the workload, but they definitely provide ways to *lighten* your workload (or at least give you better results for the work you put in). Our top recommendation here is John Bean and Dan Melzer's *Engaging Ideas: The Professor's Guide to Integrating Writing, Critical Thinking, and Active Learning in the Classroom* (2021). This is *the* evidence-based guide to mentoring writing across disciplines (including the sciences). It's an approachable guide to the history and knowledge base of writing studies and rhetoric and composition, and you can dip into it to address specific issues if you don't have time for a full read-through. It also provides exercises, examples, and extensive references for nearly every dimension of writing instruction you're likely to contemplate.

These three books only scratch the surface, of course. In the appendix we provide a longer list with more detailed annotation of the books we'd recommend you consult, own, or make available to the developing writers you mentor.

BLOGS, WEBINARS, AND THE LIKE

While books are comfortably familiar, there's a lot of similar content in digital form: blogs, webinars, podcasts, and so on. Of course, because blogs and similar media are so easy to publish (independent of the author's actual expertise), this kind of resource needs some cautious vetting. But there are high-quality examples, and you can recommend them to developing writers just as you can books—only with lower barriers to access, as many or most are freely available. We suggest some we find particularly valuable in the appendix.

WRITING EXERCISES

Just as there are stacks of writing guidebooks, there are mountains of writing exercises available for the picking. Some are in guidebooks (see the previous section and the appendix); some are in workbooks; some come in handouts from writing centers; and a bewildering profusion can be found on the internet. Here are a few ways you can sort through them.

1. *Be specific.*

You and your students will benefit more from exercises when you're clear about the writing issue or task you're working to improve. The alternative is frustration both for you (you spend too much time wading through writing exercises that won't help) and for them (they don't quite see the point in the exercises they're given).

Be aware that searching for the specific exercises you need can be a bit hit-or-miss. For example: we searched on "writing activity develop self-editing checklist" and "in-class scientific-writing activity IMRaD"; in each case, we ended up with few useful results. A search on "in-class scientific-writing activity outline" produced much better options. It can pay to ask a librarian or a subject expert (perhaps at your campus writing center) to help you refine your search. We can assure you, though: there are corners of the internet where useful writing exercises abound. They are often produced by people who both study and teach writing, usually in departments such as English or rhetoric and composition. Once you've found one or two (by web search or on social media), their blogs and websites unfold like mycorrhizal networks.[3]

2. *Consider exercises from people trained in writing instruction.*

Because most scientific-writing mentors don't receive evidence-based instruction in how to teach writing, we suggest that you resist the temptation to default to writing exercises from peers in your own field. To clarify: we're not dismissing the expertise that can develop from many years of mentoring scientific writers—after all, that's Steve's story. However, if you're stuck, frustrated, or starting out, you'll save time and stress by choosing evidence-based resources. Once you have a strong foundation in evidence-based writing instruction (perhaps gleaned from this book and other reading), you'll be better equipped to assess writing exercises made available by people in your own discipline.

How do you find these writing experts, given that they work in disciplines outside your own? Here are a few starting points:

- Look for scholars who teach first-year writing or other types of writing courses, who work in or direct writing centers or programs, or who are on the editorial boards of applied journals about writing instruction.
- Search for peer-reviewed journals that emphasize writing instruction. You'll find journals such as *Prompt: A Journal of Writing Assignments*, where you'll encounter a wide range of peer-reviewed, evidence-based writing assignments and exercises that you can borrow or modify. (More such journals are sprinkled through the references at the end of this book.)
- Reach out to your colleagues in teaching and learning centers, writing centers and programs, and English departments on your campus, or at other institutions with well-regarded programs in these areas.

Now, two important caveats. Many folks in writing studies and rhetoric and composition use jargon that might feel impenetrable to you. That's one reason we wrote this book! In addition, most of them don't specialize in scientific writing as a genre and may be unfamiliar with its norms. However, *you* have that latter expertise. With this in mind, you'll find it useful to look for exercises you can use in two ways.

1. You can *adopt* core exercises used in rhetoric and composition to (a) ensure students are on the same page, and (b) address misconceptions and

key writing concepts. Consider, for example, the TEA (*t*opic, *e*vidence, *a*nalysis) paragraph. You can demonstrate the parts of an effective paragraph and have students create or revise paragraphs to achieve the TEA structure. You may be surprised to find how powerful such a simple exercise can be—and there are many such exercises available to unlock the mysteries of writing for students who need foundational instruction.[4]

2. You can also *adapt* exercises from rhetoric and composition. Review exercises from multiple sources and look for what they have in common. Those commonalities tend to be the expert bedrock that writing instruction is built on. Retain those while still modifying the context and specificity of the exercise to suit the norms and best practices you know from your own discipline.

SYLLABI

Perhaps you love developing a course outline, or a whole curriculum, from scratch. But if you don't (or if you'd rather invest that time somewhere else), you're in luck—many instructors now post their course syllabi online, where you can reference and leverage them.[5] Our advice for filtering through these syllabi is similar to that for writing exercises. You're going to get more bang for your buck if you concentrate on syllabi from people who have an evidence-based foundation informing the way they design and deliver their courses and workshops.

Your peers and colleagues

Your peers and colleagues are obviously useful resources when they share their own writing exercises and syllabi—but that's not the limit of their utility. We suggest thinking about how you can tap into the expertise of peers both in your own discipline and outside it.

START WITH DISCIPLINARY PEERS

There may be writing studies–trained people or people with a track record of success at teaching writing in your department or in other STEM departments on your campus. (Bethann and Steve aren't the only writing nerds in North American STEM!). These folks often "speak both languages"—they understand what you and developing writers in your discipline are trying to do, *and* they have familiarity with evidence-based writing instruction. They can therefore be valuable sources of advice and can help you access more of the literature that informs effective writing instruction. You can also look for narrower kinds of expertise: folks who have particular experience with, and great advice for, efficiently reading and tracking sources, managing citations, making figures, or any of the other myriad activities that go into writing. This point about experience applies to genres, too: if you're working with grant writing, you can involve someone who's served on a review panel; if you're working with op-eds, you can involve someone who's written a few. As you narrow the expertise you're looking for, you may broaden the pool of people you can tap—so be open to recognizing the skills of a postdoc or senior grad student.

How, more specifically, can you involve these peers? They're unlikely to agree to take a stack of assessment work off your hands—but you can invite them to give a guest lecture or meet with a student or a group of your mentees. It's not just their particular expertise you can leverage, either. Sometimes writing insights gain more traction with students when they come from, or are reinforced by, someone other than a primary mentor.

ENGAGE WITH COLLEAGUES BEYOND YOUR FIELD

Beyond your department and discipline, you may find support in the sciences more broadly. But you should also search for colleagues in units where the craft of writing and evidence-based writing instruction are intrinsic: English departments, writing programs, writing centers, teaching and learning centers, and the like. The professionals in these disciplines are invaluable resources for you and your students, especially with writing issues that are fairly universal, such as plagiarism, paragraph structure, or outlining. You'll want to go into such interactions with an eye open to (1) what dimensions of their advice you can use in your discipline (it's more than you might think at first), and (2) whether you can find someone familiar with scientific writing or willing to develop that familiarity.

DON'T ASK FOR TOO MUCH

Of course, the people who can help you are also busy being scholars, instructors, and individuals with rich personal lives. Be respectful of people's time and expertise, and don't just send your students *at* them or cold-call colleagues with a substantial ask in mind. Establish relationships well in advance by engaging with relevant programming on your campus or theirs, developing a genuine interest in other disciplines' professional work, and socializing beyond your disciplinary silo.

When you do approach colleagues for support, make appropriate-scale queries. For example, don't ask for extensive writing support without offering compensation or reciprocal aid. This is particularly important if you're approaching colleagues in a humanities, fine arts, or social sciences department. These faculty are sometimes weighed down by teaching loads double those of STEM colleagues; in addition, their job descriptions (and review, tenure, or promotion processes) rarely encompass the sort of collegial service you might be requesting. For this reason, when possible, start by approaching service and consulting units (writing centers, teaching and learning centers, etc.). The colleagues you encounter there are paid to support you and your students, and thus their efforts to do so actually "count" for them.

In suggesting that you don't ask for too much, we don't mean to discourage you from asking at all. In particular, don't underestimate the value of what you can offer as reciprocal help. You might well find colleagues teaching rhetoric and composition, for example, who would love a guest lecture on norms in scientific writing. Explore!

DON'T TAKE SUPPORT FOR GRANTED

If you find value in the kinds of help we considered in the previous section, then you should do what you can to ensure that universities and other organizations hire, support, and retain the people and programs who provide it. Therefore, please advocate for the colleagues who provide writing support to you and your students. For instance:

- Actively support proposals for writing/communication courses and workshops to be offered or included in curricula—including yours.
- Advocate for such courses and workshops to count toward colleagues' regular teaching loads. Currently, they're too often taught as overload (work above what's nominally required, and thus rewarded only informally if at all).
- Support colleagues' requests for resources needed to run such courses and workshops (TA allocations, full-time positions, program funding, etc.).
- Value and encourage scholarship relating to writing and communication within your field and department.
- Advocate for writing and instructional scholarship to be appropriately counted and rewarded in promotion and tenure reviews.
- Push back against proposals (sadly, they're common) to trim or cut programs that provide writing resources at your institution.

REMEMBER: NOT ALL ADVICE IS CREATED EQUAL

Although we've said it before, it's worth repeating: some advice is good, and some is . . . less good. Whether you're looking within or outside your discipline, we insist you look for peers who have evidence-based training or deep experience in *effective* instruction and mentorship of writing.

These might be fighting words. But strongly held opinions, and even the regular delivery of academic-writing courses, do not in themselves merit the peer-as-resource status we're talking about. Strong opinions about writing are everywhere, but a remarkably large fraction of them focus on so-called "correct" writing and adherence to personal stylistic preferences, without any grounding in pertinent literature. (You've probably encountered this more than once in comments from reviewer 2.) Fewer hard-and-fast writing rules exist than most people think, even in scientific writing.[6] Giving these strong but unsupported opinions weight will just get you tangled up in unresolvable debates about whose way is better; the truth is, there are many effective ways to write anything. If your peers seem especially pedantic about grammar, punctuation, or style, look elsewhere for evidence-based advice.

Writing centers

Writing centers are one of the most useful (but least-used) resources for scientific writers in a campus environment. For those who mentor scientific writing, we'd say the same. While writing centers are calibrated to help writers (and they're great at that), their staff can also inform and enhance your approach to instruction and mentorship—even if they don't actively offer programming to support mentors. You'll get the most out of writing center support if you can be explicit about what you need and yet be open to suggestions for your growth as a writing mentor. Does this seem a bit contradictory? Let us explain.

There's a common mismatch between scientists and writing center staff that must be navigated (but that can be productive when navigated well). Few writing centers have personnel who are steeped in the specifics of scientific writing. Conversely, few scientists have engaged much with the scholarship on writing instruction. (Because you're reading this book, you're on your way to becoming an exception!) You may, therefore, have a narrower sense of the advice and resources you need than what writing-instruction experts may perceive. To make the interaction productive, do three things: (1) state what you think you need, (2) work to translate the norms of scientific writing into broader writing lingo, and (3) make clear that you're open to any feedback and resources a writing center can offer. This will help you understand what the writing center does and how they approach assisting writers, and they'll appreciate your sharing with them the disciplinary conventions and expectations of scientific writing in your field. A good conversation can lay the foundation for support arrangements that will be productive for you and feasible for them.

Beyond direct support for your own work as writing mentor, engaging with your writing center can help you understand the support available to students. The options are diverse and vary among institutions. However, you can anticipate some combination of individual coaching and group programming such as workshops; you can recommend (or even require) that your students use either.

INDIVIDUAL COACHING OR CONSULTATION

Most writing centers offer developing writers the possibility of individual meetings with writing consultants. Undergraduates are the most typical clientele, but most writing centers are available to graduate students and university employees of all types. The typical support format is a one-on-one meeting with a professional or peer consultant (the latter being an undergraduate or graduate student[7]) and will normally focus on a writer's specific writing task or product. The developing writer will be expected to arrive to an appointment with the following: (1) instructions for the assignment they were given (lab report, Methods section, thesis chapter, etc.), (2) their draft text, and (3) at least one specific request or question. This preparation is necessary because writing center consultants don't draft, revise, or provide line edits. Rather, they provide comments and advice in keeping with the thematic feedback and minimal marking approaches we discussed in detail in chapter 3. Some students are frustrated that consultants won't "fix" their writing, particularly if they're accustomed to you or other mentors doing so. However, writing centers are dedicated to a growth process that includes developing writers learning to revise their own writing.

To make this support widely available, many writing centers offer in-person, remote, and asynchronous consultations, meaning that even students with substantial course loads, work and caregiving responsibilities, or other constraints can still access writing center support. Furthermore, most writing centers allow students to book repeated consultations. Thus, there's potential for a writing center to support student writers' rapid development while substantially reducing the amount of time you spend addressing students' early drafts or writing foibles. Keep in mind, though, that such support can supplement or reduce your effort but won't *replace* it. It's your responsibility to lead the writing instruction for your students.

Getting students to seek support from writing centers may require some encouragement on your part; we've found that relatively few do it on their own. There may be several reasons for this. Some students have trust and expertise issues with consulting peer tutors. Some tutors are too specialized (especially in English and the arts) to support STEM writers effectively. And because writing centers emphasize self-directed growth and higher-level concerns like organization and clarity rather than offering line edits, students expecting edits and "correction" may be disappointed. All these concerns can be overcome with some explanation of what writing centers do, and why.[8]

We'll close with a note of caution (or really, a candid plea) from our writing center colleagues: if you're going to require your students to go to the writing center, the writing center would *very much* appreciate a heads-up and a chance to discuss what they can and can't do. Most writing centers, for example, won't have the staff to handle every student in a large class showing up just before a deadline! And requiring multiple consultations per week is a bit gauche—you might be monopolizing resources meant to be available for the entire campus. You can work with your writing center to identify a sustainable and productive balance.

WORKSHOPS AND WRITING SUPPORT PROGRAMS

In addition to one-on-ones, your writing center may offer more structured, group-focused programs. They might offer workshops on specific types of writing (a cover letter, a résumé, an abstract, a persuasive essay, a review paper), or visit your class to describe what the writing center offers or to provide guest lectures—just as a few examples.

Many writing centers also offer programming addressing writing mindsets or habits (e.g., cohort writing fellowships or the grad-level Scholarly Writing Practices program Bethann co-developed and facilitates at her university).[9] Writing centers may also offer writing events and programs, sometimes called "Shut Up and Write," "Write-In," or "Writing Boot Camp." These range from ad hoc and pop-up events to regular weekly sessions where students or writers at any career stage can gather to write alongside others (in proximity, not as coauthors). How do these programs share the workload? As we've stressed, teaching people to write is more than teaching them style or grammar. Motivating developing writers is part of mentorship. Such programs can help the writers you work with build regular, productive writing habits with less need for your direct intervention. Writing center programs may provide specific structure, writing prompts, timed writing "sprints," or goal-setting and progress-reporting systems for writers who are building their independent ability. All this can be remarkably helpful for students who struggle to write on their own or to organize their writing projects into consistent writing sessions or tasks, or who simply benefit from writing in a setting where other people are also writing.[10]

Librarians and the resources they steward

Like writing centers, librarians and the vast array of resources they steward are underused, but they can play a big role in your growth as a writing mentor and in your students' growth as writers.

Librarians have a wealth of specialist knowledge about peer-reviewed and broader resources (e.g., archival material, gray literature), including extensive materials relevant to writing instruction. For those of us steeped in our own disciplinary specialties, it can be hard to remember what it's like to delve into a discipline that's new to us without knowing its field-specific vocabulary. As you look for writing and writing pedagogy resources, the right librarian can help you find what you need—among other things, by helping you define search terms that will get you into relevant bodies of literature.

Librarians can also help developing writers directly with many topics and tasks. Just as we've suggested for you, they can help developing writers access unfamiliar bodies of literature that can inform their writing and learning. More obviously, perhaps, they can help writers learn to navigate reference databases, effectively conduct literature reviews, judge the reliability of sources, and more. Furthermore, libraries usually offer workshops and resources to coach developing writers (and mentors) on the use of reference-management software to wrangle individual writing projects and/or to manage the potentially unwieldy libraries of references that build up over one's career. All this coaching and advice can be one-on-one or in group workshops, either in person or virtual. You can usually even book a librarian to lead such workshops during your scheduled class or lab meeting, or to give a guest lecture.[11] You might invite librarians to talk about literature searches, evaluation of online information, citation practices, reference management, plagiarism, scientific publishing, data management and archiving, selecting target journals, niche research databases, and much more.

Librarians can also be helpful for finding publication outlets for your and your students' projects. They are well equipped to provide advice about identifying appropriate outlets—including both the typical outlets for academic work (major journals) and other publication options (repositories, locally published journals, books, popular press, and more). Sometimes libraries offer small grants to support development of open-access educational resources or to subsidize open-access publishing. Check with your library to find out the scope of support they offer.

Students' own peers

We'll close this chapter with one of the most obvious options for sharing your workload and providing developing writers with extra support: their own peers. We'll discuss writing groups and peer mentoring, and then, at more length, in-class peer review.

WRITING GROUPS, "NEAR-PEER" MENTORS, AND PEER COAUTHORS

There are several ways to set up pairs or groups of developing writers to learn together and from each other. Most simply, you can encourage (or require) the writers you mentor to develop and participate in writing groups. Such writing groups can be structured in many ways, usually informed by the ultimate goals of the group.[12] If the point is to encourage regular and productive writing habits, the group might focus on accountability and progress reports, or simply on members writing alongside other people (the approach used by "shut up and write" groups[13]). If the goal is for writers to improve each others' products, the group might use an organized feedback or workshop model where people receive feedback from partners or the group on a piece of writing. Even the act of deciding what they need can be a productive way for a group of writers to grow in responsibility for their own development.

If we stretch peers to "near peers," more advanced developing writers (such as senior students or postdocs) can help mentor earlier-stage writers.[14] These relationships may be most productive when they're made explicit, when developing writers are expected to engage actively, and when near-peer mentors are supported and rewarded for their contributions. That last point might mean you take responsibility for ensuring that near-peer mentors have access to coaching, advice, and reasonable release from other duties so they can actively support their mentee(s). Otherwise, you're likely to see uneven feedback and, quite possibly, resentment.

Finally, you can encourage coauthorship among the developing writers you mentor (as science becomes ever more collaborative, this is likely to happen regardless). Building on the idea of near peers, deliberately setting up collaborations involving more- and less-advanced developing writers is likely to pay off the best. In many cases it's even possible for graduate students to coauthor thesis chapters with each other.[15]

IN-CLASS PEER REVIEW

There's extensive literature on effective peer feedback and peer intervention in classroom settings.[16] We use the term "peer review" here because that's the dominant label for this instructional approach in the writing studies and rhetoric and composition literature; it is, of course, distinct from the act of refereeing literature in one's discipline.[17] In-class (or course-based) peer review is an appealing framework that has been adopted extensively in writing classrooms as a way to reduce instructor workload while increasing feedback to students. It's not just that, though; pedagogical benefits of peer review arise because it's a way of teaching-to-learn and co-constructing knowledge, a richly metacognitive activity for students.[18]

We're not suggesting that peer review is a panacea. Both the literature and hallway conversations identify downsides and limitations when students give feedback to their peers.[19] For one thing, many students resist giving or receiving peer feedback on writing. They push back for valid reasons, such as receiving unproductive feedback, putting in way more effort than their peer-review partners, or feeling poorly equipped to provide feedback. Students who are conditioned to look for a single "right way" to write can also resent it when they perceive mentors to be off-loading teaching (or as they think of it, "correcting") onto other students, and also when peers don't or can't provide "accurate" corrections.

Fortunately, students do want, and often solicit, extra feedback on their writing—they just want it from people they individually trust[20] and who they believe understand the challenges they face as developing writers. They seek this feedback from friends, family, or classmates they perceive to be "good" writers or knowledgeable about "good writing." We used some scare quotes there because this intersects with another issue with peer review: just like some mentors, students can misperceive what good writing is. Frequently, they advise their peers to use third-person, passive-voice writing and impenetrable, multisyllabic words in order to "sound smart" or science-y. But these aren't rules; they're (unfortunate) choices about style (see chapter 1). A good way to mitigate this problem is to take direct aim at the "correction" fallacy: make clear to students that there are few rules in writing, that as peers they are making and receiving *suggestions*, and that you will facilitate follow-up discussion between peer reviewers and reviewees.

Because students solicit and value feedback from people they trust or perceive to have expertise, you *can* use peer review to provide support for your students. Some tips:

1. **Avoid creating circumstances that consistently lead to negative peer-review experiences.** Here are a few easy ones: Don't merely assign peer review and expect it to be meaningful. Don't just tell students peer review is important and assume they'll believe you and experience it as such. And don't assume that students' goals for peer review are the same as yours. Instead, actively frame and facilitate peer review by following the next several tips.
2. **Articulate your goals to yourself.** Why are you assigning peer review?[21] Would you like peers to help with topic selection, transitions, introductions, conclusions, titles, frames for a piece of writing, or something else? Do you want students to discuss, with authentic readers, the readers' experience with a draft? Do you want peer review to provide experience with co-constructing knowledge, to identify next steps for revision, or to help students expand their perspectives beyond the intentions of the writer to the needs of the reader? Will peers provide polishing line edits at the end of a writing project? Or perhaps you'll have multiple rounds of peer review with different goals. Whatever the case, the clearer you are on your purpose(s), the more readily you'll design productive peer-review experiences for your students.
3. **Articulate your goals to your students.** Tell developing writers not just what you want them to do, but what you want them to get out of doing it. You can provide rubrics (see chapter 4), examples of your own feedback, and concrete examples of productive peer review (or writing feedback more generally) for the type of assignment and the stage of peer review you're intending. Talk through these examples with students so it's clear what kinds of feedback you're encouraging them to provide and how they can provide it. Without this step, you risk confusion and mutual frustration, because student goals may be misaligned with yours.
4. **Discuss past peer-review experiences.** Encourage students to acknowledge any benefits of peer review that they've experienced (even if it's a short list). Make room, too, for students to air their concerns about, or resistance to, peer review. Such discussion can build trust and shared goals among students and can alert you to ways you may need to troubleshoot the process. Circle back to this discussion regularly if you use repeated rounds of peer review with a cohort of students.

5. **Spark discussion of how peer review helps writers.** You can prime students for peer review by assigning essays, podcasts, and the like in which people discuss the importance of feedback to their writing. Engaging with other writers' perspectives on getting feedback and revising will set up the idea of writing as a process that benefits from interaction with other people. As a bonus, you're foregrounding the more general idea of writing as a craft and a process, not merely as an end-product text (see chapters 1 and 6).
6. **Facilitate discussion in which students identify what kind of feedback they'd value, and when in their writing process they'd value it.** If you've already conducted the previous steps, students will be able to engage actively in this discussion, which makes the value of feedback more concrete for their own writing. As a result, you'll be able to connect students' goals to the peer review you propose for your class—setting up students to invest more productively in enhancing each others' writing.
7. **Remember that students solicit and value feedback from people they trust.** Achieving your goals for student peer review hinges upon cultivating a community of trust and collective learning in your classroom. That's a broader set of pedagogical efforts than we can tackle in this book, but great resources do exist.[22]

With these steps sorted out, you can implement a range of peer-review models in your classroom (or lab group), with students understanding that you're doing it to benefit *them*, not just to lighten your own workload. (See more about building peer review into lab groups in chapter 10.)

There are lots of practical details about implementing peer review. You can organize it as a scheduled activity during class time or make it asynchronous (most learning-management systems provide tools for assigning and receiving peer review online). If you involve more junior mentors such as TAs, you'll need to arrange for some training and support before they can effectively guide the kind of peer review you want. Finally: consider whether peer review is the last step or an intermediate one before review by a TA or by you (in a "hierarchical" feedback scheme[23]). Students may be more comfortable with peer review if they also receive *your* feedback, because they know that your review has a close relationship to assessment (and they assume that your feedback is more informed than peer advice).

We'll close this section with a nudge to do some more reading, because providing specific, detailed instructions on how to facilitate peer review for a given course context is beyond the scope of this book. There's extensive literature that considers a wide variety of teaching situations.[24]

Resources for things you should *not* try to do alone

There's one more category of collective resources that will be relevant—and indeed, necessary—for many of your students. Most campuses provide a range of services and programs to support students who are dealing with health issues, mental health conditions and challenges, disabilities, etc.[25] They will also provide interest- and demographic-group support (e.g., international student groups, identity affinity groups).We have two central recommendations about these resources: (1) anticipate that many of your students will need them (at undergraduate and graduate levels) and (2) take your students' needs for them seriously. To help your students benefit from these resources, consider the following:

1. Most universities tell incoming students that student support and student success are directly correlated: students who seek out support do better. They're right. So, don't look at support for students as something to resort to. Instead, encourage students to proactively pursue all the avenues of support available to them.
2. In order to accomplish that last point, we recommend that you familiarize yourself with the support services that are available to your students. Make these services evident by displaying posters and/or providing a resource list as part of your syllabus or other course materials.
3. Reach out to your campus's office of disability support services or equivalent. Ask them to point you to resources or to explain in detail what your responsibilities are as an instructor and advisor. Ask them for detailed information about applicable laws governing students' rights to accommodations. If you mentor, advise, or serve on committees for graduate students, take special care to verify what types of accommodations the office can facilitate as well as what additional types of accommodations you're allowed to facilitate.[26] Typically, mentors are allowed to provide more accommodations than most faculty realize.
4. Some campuses also provide free legal services, housing support and housing dispute mediation, ombuds offices, and even (though rarely) financial support for crisis and emergency situations. While such services may be less widely available than health support, it's worth seeking these out, too, so that you can make your students aware of them *before* they need them.
5. Similarly, many communities offer support services that universities may not, including free or reduced meals, housing assistance, free winter clothing (especially vital for many international students), free or very affordable health care (most relevant in the United States and if a

student has a partner or dependents who aren't covered by the university's student health services), support for victims of abuse and domestic violence, etc. Community connections are vital for most people's well-being,[27] and students will benefit from you pointing out ways to connect with people off-campus, too (e.g., library programs, book clubs, interest groups sponsored through local businesses or nonprofits).

You'll notice that we haven't suggested that you engage much with these types of issues yourself. If you're a faculty advisor or mentor for one of the support services or social activities available on your campus or in your community, you might be slightly more involved. However, if you're not a trained therapist, health professional, lawyer, or the like, it's not appropriate and may not be legal for you to intervene in your students' issues, even with the best of intentions. Be considerate of your students' challenges, but do not attempt to diagnose or treat them yourself.

By the way: perhaps you're thinking that most of this section doesn't have much directly to do with writing. You're right, in a way—but remember that writing is done by people (your students), and people need all kinds of support. Writing is pretty hard to do when people are struggling and need support for more fundamental aspects of their lives. If you connect your writing students with resources that help them not just with writing but more broadly in their learning journeys, we think you'll agree that you'll have done a good thing.

Parting thoughts

While this is, we admit, a long chapter, we haven't exhausted the list of places you can look for partners to help students write. At least, though, we've shown you that you don't have to shoulder the burden alone. Be creative and proactive in the support you seek out. You're bound to build your capacity as you work to support developing writers.

8: Teaching and Mentoring the EAL Writer

What kind of challenges are faced by people writing in English as an additional language (EAL)? • What kind of useful advice can you give the EAL writers you mentor? • Is plagiarism an issue of special importance in mentoring EAL writers?

A striking thing about modern scientific literature is that nearly all of it is produced by writers working in a language that isn't their native one. Over 90 percent of twenty-first-century scientific literature is published in English;[1] and yet, worldwide, most scientists grow up speaking and writing some other language first.

In one way, a system with a single dominant language makes sense: the adoption of a common language (whatever it might be) means that science can function worldwide with everyone needing to learn at most one additional language. (It wasn't always English; Latin, German, and French have all had their day). The result is an efficiency without which science couldn't be as global as it is. That efficiency, however, accrues to science as a whole, while the costs are distributed inequitably. In particular, English-as-first-language speakers are privileged while speakers of English as an additional language (EAL) pay the costs.[2]

Why does this matter in a book about teaching and mentoring developing writers? Because the growing internationalization of science means that if you haven't mentored an EAL writer yet, you surely will soon—most likely, many of them. EAL writers face many of the same challenges as English-as-first-language writers, and most of the techniques we've discussed so far are applicable to mentoring them. However, they also face additional challenges—and crucially, so do their mentors.

Before we dive in, we'll acknowledge that in one short chapter, we can only scratch the surface of mentoring EAL writers. You should see this chapter as an orientation or a jumping-off point, and we encourage you to dig deeper into the resources we suggest. (While this is true of every chapter in this book, it might be most true of this one.)

Start with empathy

We'll start with something that probably sounds both trivial and obvious—but we think it's important. EAL writers will greatly appreciate a simple acknowledgment of the burden they're shouldering. There's abundant literature documenting that most EAL writers feel disadvantaged compared with their English-as-first-language colleagues.[3] Such writers may feel culturally isolated in other ways, too, especially when they're living and working away from home. Some empathy from their mentors will at least help students recognize that their feelings are valid and that other people care. Empathy, coupled with a commitment of support, can help shift an EAL writer from feeling excluded to feeling entitled to participate in the community of science.[4] Of course, if you're an EAL writer yourself, you can take this one step further by being open about your own learning journey, sharing your struggles as well as your successes.

While any expression of empathy will likely be appreciated, we strongly suggest fine-tuning it with a bit of knowledge about your EAL students' language backgrounds. That's because EAL writers aren't a homogeneous bloc. There are over 7,000 living human languages, and we certainly aspire to a world in which a speaker of any of them can find a place in the scientific community. Some EAL writers will come from countries where English is an official language widely used in education, business, and government (such as Nigeria or India). Some will come from countries where English isn't used officially but is nonetheless widely spoken (Iceland, Switzerland). For still others, English will be more culturally novel. Some EAL writers will speak a native language with some similarity to English in grammar or vocabulary (Dutch, French), while others will have a native language that works very differently (Japanese, Swahili). And, of course, each individual will have their own history with learning English: perhaps they have extensive training in conversational English but none in scientific English; or perhaps the reverse. You get the picture—it's complicated, so ask your students for their specific context. When you do, though, keep in mind that they are experts in their *personal* experience—it's best not to expect students to represent their culture, country, language, or the EAL experience overall. Doing so risks stereotyping and misunderstanding students' experiences.

Language, writing, and thinking

When you think about the task facing EAL writers, you might begin with the idea that they lack facility with writing in English, and so need to learn English grammar and vocabulary. That's certainly part of the task, but things are more complex and more interesting than that. That's because language background is at least correlated[5] with cultural and cognitive differences, and as a result of those differences, EAL writers may approach text in a very different way from English-as-first-language speakers. On top of that, of course, one EAL writer may approach text in a very different way from another.

There's a good example of the connection between language and cognition in the contrast between the right-branching structure of English and the left-branching structure of Japanese and Korean (among other languages). By "right-branching" we mean that English places the "head," or key subject word, of a sentence or phrase at the beginning, with subsidiary information following to the right: *the dog that I saw last week at the park.* A left-branching language does the opposite, with a structure like *the last week I saw at the park dog.* A reader decoding a left-branching phrase must remember a bunch of leading words without (yet) understanding them, because they modify or expand on something that isn't apparent until the end of the phrase. Likely as a consequence of their linguistic experience, speakers of Japanese and Korean show greater ability to remember the first item in a list, while speakers of English and other right-branching languages remember last items better.[6] This has implications for structuring paragraphs and other passages of text: speakers of left-branching languages may choose to place the most important material at the beginning of a passage, while speakers of right-branching languages (like English) may choose to place it at the end. Each choice is intended to meet reader expectations by placing critical information where readers will tend to remember it easily[7]—but linguistic variation means that there's no single strategy that will accomplish that for all readers.

There are many other examples of these linguistic mismatches (some of which we'll touch on below). Because of them, EAL writers may sometimes need to strain, or even violate, stylistic or discourse conventions from their native language in order to produce text that's recognizable as conventional scientific writing in English. An EAL speaker isn't just asked to learn English vocabulary and sentence structure; they are to some extent being asked to learn English *thinking.* At best, learning to think in another language expands a developing writer's options; at worst, it's forced assimilation into the prestige dialect of Standard Academic English (see chapters 1 and 3). As a mentor, you can't be expected to be an expert in comparative linguistics and cognition. But being aware of some of the features of the languages your students speak can help you reduce their confusion and frustration—and yours.

Some common EAL bugbears

Because EAL writers aren't all the same—even those sharing a language background—it really isn't possible to provide a list of the English-writing issues that trip them up.[8] Nonetheless, for two reasons it's useful to examine a few of the more common issues. First, precisely because they're common, you'll likely work with developing writers who are troubled by them. Second, if English is your first language, these examples should clarify that there are features of English writing that seem completely obvious to you but not at all obvious to your EAL students. EAL writers make these errors not because they're careless or slow to learn, but because features of the English language are either extremely puzzling or extremely divergent from their first-language experience. A quick reminder, though, that many of these are minor issues best ignored until quite late in the development of a draft (see chapter 3). That's especially true if you'd like to bolster rather than shred a student's writing confidence.

- **English makes heavy use of articles (*the* and *a*/*an*); many languages don't.** Some even lack them entirely (including Chinese, Japanese, Finnish, and Farsi). Speakers of those languages often struggle with English articles. English-as-first-language speakers, who use articles intuitively, are sometimes surprised by this. But in fact, the rules governing the use of articles in English are astonishingly complex.[9] (As one bizarre example, we use articles with the names of rivers and oceans, but not lakes or bays). Most English-as-first-language speakers easily recognize "incorrect" use of articles but have great difficulty explaining why a particular usage is incorrect or what the typical English pattern is.
- **In English, simple sentences usually consist of subject, verb, and object—in that order ("I read the book"). Other languages use different orders.** Japanese and Turkish, for example, use subject-object-verb ("I the book read"); Arabic and Tagalog use verb-subject-object ("Read I the book"). As sentences get more complex, there are more possibilities. In English, adjectives usually precede nouns, while in French and Spanish they (mostly) follow them; and the right-branching and left-branching structures discussed in the previous section are more complicated examples.
- **Languages have complex systems of count and noncount nouns.** A count noun has different singular and plural forms: in English, *a dog, two dogs*; *a goose, two geese*. In contrast, a noncount (or mass) noun has only one form, being treated as an undifferentiated collective (*some money*,

more money). Other nouns can be count or noncount depending on their interpretation (two fish of one species: *two fish*; but two species of fish: *two fishes*). The rules determining countability are complicated,[10] but what's worse is that these rules don't match those of other languages. "Equipment," for example, is a noncount noun in English (*some equipment*, *more equipment*) but a count noun in French and German (*un équipment*, *des équipments*; *die Ausrüstung*, *die Ausrüstungen*).

- **English is a "writer-responsible" language, in which the primary responsibility for clear communication lies with the writer.** Japanese, Chinese, Thai, and Finnish (for instance) are "reader-responsible" languages, in which both writers and readers expect the latter to do the work of acquiring background knowledge, inferring connections, and understanding reasoning.[11] Thus, EAL writers from reader-responsible backgrounds may find the clear, linear exposition and meta-discourse typical of English scientific writing to be unsophisticated or childish.
- **Cultures differ in rhetorical approaches to argument.** English writers tend to use linear and general-to-specific structures (think the narrowing focus of the Introduction in most scientific papers), but EAL writers may have different first-language conventions.[12] Writers from Hindi, Russian, German, and Romance-language backgrounds tend to digressions and parenthetical amplification;[13] those from Chinese, Korean, and Arabic backgrounds often use a more indirect rhetorical style with many parallel arguments or multiple approaches to a subject.
- **Languages—or at least cultural traditions linked to language—favor different argumentative stances.** For example, Japanese speakers may prefer a deferential attitude to other scholars, especially more senior ones, and may avoid framing claims in ways they see as confrontational. This may mean reluctance to identify knowledge gaps or to emphasize conflict between their work and previous research—and yet, these are standard rhetorical devices (or "moves") in English-language Introductions and Discussions, respectively. Similarly, Korean speakers and Sudanese speakers of Arabic may avoid "boosting" language that they feel is inappropriately emphatic. Mexican speakers of Spanish and Hong Kong speakers of Chinese may, in contrast, use strong rhetoric that English speakers think exaggerates claims.[14]

These half-dozen examples don't exhaust the list of ways language background can color writing and either produce mistakes or lead a writer to deviate from English conventions. The point really isn't to list all those ways—nobody could make that list or remember it. Instead, the point is that as a mentor, you may be frustrated with a student's writing pattern—a mistake they make over and over again—and suspect ignorance or noncompliance with your feedback. We hope you'll now be less frustrated, knowing that there's a good reason why a student's particular language background makes it challenging for them to master a particular convention of English scientific writing.[15] You can also use this awareness to open a conversation about the writing pattern and why the English convention is what it is. Of course, you may need to do some work to figure out that "why" rather than relying only on the intuition that comes with native fluency.

Advice to give the EAL writer

If you're mentoring a developing writer who speaks English as an additional language, you'll likely find yourself asked for advice, or wanting to give it. We have good news of two sorts. First, you needn't shoulder the mentoring job alone; there are tools and resources you can steer an EAL writer to (and we'll recommend some in the next section). Second, there's useful advice you can give that can significantly ease the EAL writer's job. What follows is a list of tips—not for you, but for you to pass on to the EAL writers you're mentoring.[16]

- **Keep the text simple.** Scientific writing is notorious for long, complex sentences and paragraphs. However, the more complicated the text, the more difficult it is to write effectively. EAL writers can make their jobs easier by writing short, simple sentences. Actually, *any* writer can make their job easier this way and make their reader's job easier, too—definitely a win-win.
- **Practice.** In mastering a language, there's little substitute for frequent practice. EAL writers should therefore be encouraged to write (and speak, listen, and read) in English as much as they can. In a classroom setting, this might mean frequent, low-stakes assignments. In a graduate setting, it might mean encouraging writing projects outside the thesis—perhaps the first draft of a collaborative paper, a scicomm piece, or a grant proposal or report. More broadly, practice with *any* kind of English writing will help, so encourage an EAL writer to try their hand at blogging, newsletter articles, even letters and email. Likewise, soaking up English-language media of all kinds will help: books, comics, radio, films and TV shows are all important ways to condition a language learner's brain to the habits and idiosyncrasies of English. Such exposure matters because many language patterns aren't "logical" and thus can't be

"rationally" learned but rather must be memorized or assimilated through immersion. In general, it's best for EAL writers to engage with text (and speakers) a little beyond their current level of English competence. While this may occasionally be uncomfortable, it's how people learn languages.[17]

- **Use existing text as a model.** A common strategy for EAL writers is to find a piece of existing writing and model (reproduce in broad features) its structure and phrasing.[18] This can be an effective strategy, especially when used at the largest scale (paragraphs-to-sections) or the finest one (individual phrases). There's an obvious danger, though: when modeling shades into close imitation, the result can be plagiarism. A few words or a phrase ("was significantly correlated with") can be borrowed freely, but a succession of them can't be (and note that use of a thesaurus to swap words out one by one is neither appropriate nor effective). In practice, many developing writers struggle to find the line between modeling and plagiarism, and the issue can be more acute for EAL writers (see "Plagiarism and the EAL writer," below). There's a further danger of mimicking the less desirable aspects of scientific writing (e.g., overuse of the passive voice in service of supposed neutrality, excessively complex syntax and vocabulary).

  There's a role for you as the mentor here: you can encourage use of model texts, but also offer to provide guidance on the results. And while you can encourage your EAL students to keep track of relevant (and good) chunks of writing that can be modeled, you can also provide them. This will be particularly helpful for writing that suits common rhetorical moves like the presentation of statistical results or the hedging of conclusions.[19]
- **Use phrasing references.** EAL writers can take advantage of compilations of common phrases and structures in scientific English.[20] Such compilations are available online or as part of guidebooks aimed at EAL writers.[21] They may, for example, suggest constructions that express varying degrees of certainty or doubt (X is *likely/probable/possible*) or offer alternative ways to describe a difference between treatments (*major/clear/distinct/minor/slight difference*). Most such resources go well beyond these simple examples to more sophisticated English constructions.
- **Experiment with writing assistance software.** Software such as Word's grammar checker or Grammarly can help any writer, of course. But such software may be especially helpful for EAL writers because current versions grapple reasonably well (though not infallibly!) with

the peculiarities of English grammar and composition. Specialist software, designed expressly for EAL writers and focused more on rhetorical structure than grammar, includes SciPo-Farmácia and SWAN.[22] Our recommendation here extends even to large language models (LLMs) such as ChatGPT. These are quite good at producing text conforming to the stylistic conventions of English and of a specified genre, and EAL writers report success in asking them to revise and polish draft text for English fluency.[23] Large language models do this, by the way, in part by automating the kind of corpus analysis mentioned in the previous bullet: they mine existing writing for common constructions and deploy them when generating text. However, if asked to produce a draft rather than polish one, language models frequently invent information, misrepresent knowledge, or otherwise betray that they're only good at *sounding* clever, not at *being* clever. As a mentor, then, you can encourage appropriate use of LLMs; but you should be careful to provide guidance about what that is. Box 8.1 presents a commentary from an EAL PhD student who has found a good way to integrate ChatGPT in his writing process—notably, with plenty of his supervision. (See chapter 9 for further discussion of the pros and cons of LLMs.)

- **Avoid translation.** Some EAL developing writers report writing drafts in their first language and then translating the nearly final product into English—using either free tools like Google Translate or professional translation services. While this might seem like an effective strategy, those who pursue it are very often dissatisfied with the results.[24] There are several reasons why this may be so. Free translation tools work well only for some source languages. Professional translation can be expensive, and few translators have the disciplinary knowledge to translate scientific texts well. Perhaps just as important, because languages vary in structure and rhetoric (see "Some common EAL bugbears," above), the translated text can retain patterns from the original language and be seen by reviewers (and even the EAL writer!) as low quality. Finally, write-then-translate isn't an effective way to master writing in English in the longer term. Tools such as Google Translate can be very helpful when used to suggest words or simple phrases, but we recommend against translation at larger scales.

Where to steer an EAL writer for further help

Even with the suggestions from the last section, you may feel incompletely equipped to advise your EAL students. Fortunately, it's not all on you. You

BOX 8.1

An EAL PhD student's take on ChatGPT

Like many scientists, I invest a significant amount of time in writing. However, as a nonnative [English] speaker, I often face challenges in maintaining conciseness during the initial stages of my writing. To solve such a problem, I started using ChatGPT. Since I have started using it, I have been able to speed up the initial phases of my writing by providing informal language as an input and receiving more concise writing as an output. I actually never use it to create text from scratch when I do scientific writing; I integrate the parts of ChatGPT's output that I like and then repeat the process. Once I reach the point that ChatGPT makes my writing worse, it means I have a good piece of writing!

When my language is concise, with ChatGPT's help, it becomes easier for me to comprehend how my arguments are interwoven. This newfound efficiency has enabled me to complete my writing projects more swiftly, thereby freeing up valuable time for me to engage in additional writing endeavors—such as this commentary, also written with the help of ChatGPT.

Emanuele Giacomuzzo, University of Zurich

---

can also help them discover and access other resources, both formal and informal. For example:

- **Coauthors and friendly reviewers.** EAL writers generally appreciate—and, in fact, rate very highly—language assistance from their coauthors, or from friends and colleagues who are willing to read drafts.[25] It's worth encouraging your EAL students to seek this kind of help, and it's worth facilitating access to it. But it's also worth making sure they realize that substantial writing help with a manuscript isn't a small ask. They should thus do everything possible to make the job easier, thank those who help, offer to reciprocate, and (with your guidance) consider what degree of help might warrant coauthorship.
- **EAL writing groups.** The EAL developing writers you work with are unlikely to be the only ones at their institution. You can suggest that they start, or join, an EAL writing group. Meeting regularly can give them a supportive environment to exchange drafts, practice spoken English, or host presentations from EAL teaching or writing experts. Such groups can also help students realize they are not alone, either in cultural/linguistic differences or in their struggles as developing EAL writers. The latter awareness can be crucial for EAL writers in persisting in their

degree programs and building confidence as scientists. You can help by sponsoring the writing group if that's needed, or by helping them locate experts to invite. Writing groups with writers of different language backgrounds may be more productive, when possible: they make practicing English more natural, and compel writers to compare different perspectives on writing (and everything else).

- **Workshops and courses.** Many institutions run specialized workshops and courses for EAL writers,[26] but it's not uncommon for these to be poorly advertised. Campus writing centers might also provide EAL-specialist consultation or other services. Find out what's available where you are and make it easy for your students to attend. Similar workshops may be available online, either free or at reasonable cost. We note, though, that it's worth asking about the credentials of those offering the course. Are they EAL experts? Disciplinary experts? A mix?[27] Developing writers are sometimes frustrated by sessions offered by those who lack deep appreciation for the particular expectations of scientific writing. Finally, if this kind of service is not available at your institution, you can advocate for it becoming a priority.
- **Professional editing services.** Many EAL writers—especially those working outside the English-speaking world—report using professional editing services.[28] This may sometimes be a good use of financial resources, and some institutions make funds available to support it. However, you should advise developing writers to evaluate the credentials of an editor carefully, with respect to both language in general and the subject discipline. Furthermore, if the developing writer doesn't take time to think carefully about an editor's corrections and learn from them, any benefit will be limited to the current piece of writing. (We further discuss using an editor in chapter 9, and we addressed revision for writing independence in chapter 6.)

Plagiarism and the EAL writer

Many academics spend far more time than they want to dealing with cases of plagiarism (and even infrequent cases are unwelcome). While this is definitely an issue that applies to guiding all developing writers, it's widely supposed that the incidence of plagiarism is higher among EAL writers. This supposition may be at least partly unfair—one manifestation of bias against those not from the linguistic majority—but there's at least some empirical evidence that EAL writers do plagiarize more often.[29] It's worth thinking a bit about some of the reasons.

First, EAL writers may experience more stress about their ability to produce competent scientific writing simply because of their lesser facility with the task. If all of us are tempted to cheat a little here and there in life—and we are—then it's not surprising that EAL writers might give in to this particular temptation. This isn't an excuse for wholesale copying, of course, but it's a reason for empathy.

Second, the strategy of using model texts (see "Advice to give the EAL writer," above) is powerful, but a little dangerous. It's easy to slip up, and it's easy to think a quick pass through an automated plagiarism checker is sufficient to diagnose those slips. Clear advice about this, and an offer to occasionally check old and new text side by side, will help.

Third, and most important, writers of different cultural backgrounds may come to science with different attitudes about the reuse of other writers' text. The example most often cited is that of East Asian cultures in which respect for authority makes it disrespectful to rephrase the words of a more senior figure (or, along more moderate lines, not a serious infraction if you don't rephrase).[30] It might be worth taking time to help EAL writers learn conventions of direct quotation in English, and in scientific (or other genre) writing in particular. Even though quotations are used infrequently in scientific writing, a direct quotation can function as a transparent, intermediate stage toward rephrasing source texts in one's own words.

We hope it's obvious that we aren't saying that EAL writers are less ethical than those who grew up speaking English. Instead, we're suggesting that mentoring EAL writers may involve greater attention to teaching about plagiarism, and that this teaching needs to be grounded in cultural sensitivity.

A final word

We'd like to end this chapter where we began it: with an acknowledgment that EAL scientific writers are doing something remarkable and overcoming great challenges to do it. Your support, and your empathy, will make a big difference.

9: From Pencils to ChatGPT

TOOLS TO IMPROVE WRITING, AND WRITERS

What kinds of tools are available to writers, and what's the difference between a process tool and a text tool? • How can you guide the writers you mentor to use tools of each type effectively? • What about ChatGPT (and related "AI" tools)? Can you guide students toward using AI tools effectively and ethically?

Once, we thought humans were the only tool-using animal. We now know that's not true—elephants, octopuses, and several species of ants use tools, among others—but tool use is still something deeply ingrained in us as a species. Each of us has a favorite hammer or chef's knife or set of pruning shears. We're constantly inventing new tools, too. Some of them are useless gimmicks, but some make enormous improvements to the efficiency of our tasks and the quality of our lives.

Writers use tools, of course. Some tools help with the *process of writing*: pencils, keyboards, Pomodoro timers. Other tools help with *what's being written*: spellcheckers, reading-level analysis software, and yes, ChatGPT. We'll call these "process tools" and "text tools," and deal with each in turn. The central question: how can we help developing writers use these tools in ways that complement and build their skills?

What tools are for, and three key questions

Tools can do different things for their users, and we think this realization is helpful for the writing mentor. Some tools simply reduce drudgery by taking over an uninteresting task (reference-management software for reformatting citations is an outstanding example). Others let a user accomplish a task they simply couldn't otherwise (Google Translate, for all practical purposes; any of us can learn additional languages, but nobody can learn *all* languages). Still others allow a task to be completed by a user with less skill.

In an instructional context, it's this last category we're usually most worried about. What if writers become so dependent on grammar checkers that they can no longer construct a coherent piece of text without one?

If you're considering whether, or how, to use a tool with the developing writers you're mentoring, we suggest beginning with three key questions. The answers will shape both the way you incorporate the tool into mentoring and teaching, and the advice you give your students about its use.

First: are your students likely to use the tool no matter what you do? Unless you're willing to require that writing happens with pencil on paper, right under your nose, nothing will prevent students from using spellcheckers—just as King Canute's order failed to stop the tide from coming in.[1] For writing tools that fall in this category (most, we suspect), you can move rapidly on to the question of *how* to use them.

Second, what can the tool do well, and what does it do poorly? This isn't always obvious to a developing writer, and you can make a big difference by suggesting the appropriate use of each tool.

Third, how can you ensure that use of the tool supports the development of a writer, not just of a piece of writing? While there's nothing wrong with a tool that helps polish some writing, it's far better if its use helps a developing writer write better next time (perhaps because they think carefully about what it *does* to polish the writing).

Once you've asked yourself these questions, you can decide whether you'll encourage students to use a particular tool, whether you'll *allow* them to. Is it OK to have Uncle Ted check their spelling? To run a draft through Grammarly or ChatGPT? To pay a professional editor?[2] And if students do use a tool, should they make that use visible to you (and how)?

Process tools

Process tools are those that primarily help writers manage the physical and mental aspects of writing efficiently—like a comfortable keyboard or the mood music that helps someone focus. Most of us have few, if any, pedagogical reservations about tools of this sort; we struggle to think of any context in which using a word processor (for example) could be thought of as cheating, or as a "crutch" writers use to avoid getting better at the actual task.

As a mentor, your primary interest will be in telling your students that process tools exist and encouraging them to experiment with them. That experimentation is important because all writers have different struggles—some of us procrastinate terribly; some start the job but are easily distracted;

some have difficulty with the physical task of typing for long periods. We can't review this kind of tool exhaustively, but here are some that are available. We're consistently surprised how few writers (developing or otherwise) know about them and take advantage.

- **Focus assistants.** Digital distraction is the bane of many a writer. It may seem simple just to resolve not to open Facebook or TikTok or *The Guardian*, but many of us are very bad at sticking to such resolutions. Browser extensions and other apps are available to help enforce focus on writing by (temporarily) blocking alternatives: StayFocusd (for Chrome), Leechblock NG (for most browsers), Freedom (which blocks both websites and apps), and more.[3]
- **Pomodoro timers and apps.** The Pomodoro technique is deceptively simple but works for many writers: you resolve to focus completely on a task for 25 minutes, then take a 5-minute break, then repeat. (You can of course experiment with different time blocks.) All that's needed is a timer, ideally with two timing channels; hundreds of apps and devices provide that. One can couple the timer with a focus assistant, although there's something about the clearly defined but short focus period that makes anything else unnecessary for many.
- **Notetaking systems.** Pencils and index cards are still around, but software can now expand on their capabilities. Apps like Evernote, Capacities, or Obsidian allow users to make notes, link them, reorganize them, connect them to scheduled tasks, and more.[4]
- **Project-management apps.** Many developing writers struggle with knowing how to approach, begin, and work through a large project like a scientific paper or a thesis. (Or a book!) Apps like Asana, Ayoa, or Goblin Tools are designed to help break large projects down into manageable chunks, and then to schedule those chunks and track one's progress through them.[5]
- **Word processors.** Yes, we know, everyone has Pages, Microsoft Word, or a free alternative such as LibreOffice Writer or Google Docs. These are familiar, powerful, and easy to use across teams of writers. But many writers aren't familiar with their full suites of features; for example, most of these tools will now read text aloud, which is extremely helpful in catching problems during revision. Furthermore, there are software tools that work differently. Scrivener, for example, combines the ability to write a document with tools for notetaking, organization (of both notes and complex documents), task scheduling and goal tracking, and more.[6]

- **Collaborative writing systems.** Most writing in science is collaborative: co-written, or at least reviewed and edited by multiple people. You can help developing writers see co-writing not as a dreaded "group project" but as an expected and productive aspect of professional work. While you're probably familiar with Word's Track Changes feature, in our experience even our own collaborators aren't reliably familiar with the equivalent functions in Google Docs, let alone other tools such as Dropbox Paper.[7] Thus, we recommend you embed the learning and use of these programs and functions, *as part of co-writing,* into your mentoring. You'll help students further by exposing them to tools that support different styles of collaborative writing. When two writers work sequentially—for example, one writing a complete draft and then another working to revise it—they need only Word's Track Changes or the equivalent (plus some careful attention to version control). On the other hand, when two writers work simultaneously—contributing at the same time to a document as it grows and changes—then a tool designed for that purpose (like Google Docs, Dropbox Paper, or the online implementation of Word) will be more suitable.
- **Dictation software.** Many developing writers struggle to get text on a page. For most, it's not the physical act of typing (although dictation does play an important role in accessibility). Rather, the issue is that a blank page is intimidating, and the slow pace at which key pecks replace blankness with text can be disheartening. Dictation is a valuable workaround: many developing writers find it easier to *talk* a passage of text than to *write* it.[8] Not long ago, dictation was the domain of expensive, specialized software;[9] but now, dictation (or "speech to text") capability is available just about anywhere. Windows, Apple, and Android operating systems all include speech-to-text tools, and Google Docs, Pages, and Word have dictation options.[10]
- **Reference managers.** Few writing tasks are as tedious as reformatting a list of cited literature to conform with the requirements of a target journal. Fortunately, a software reference manager can (nearly) automate this task. It can also maintain, curate, and search one's own library of PDFs and other sources. In many cases, these tools integrate with a word processor to manage in-text citations. EndNote, Mendeley, and Zotero are among the most widespread options as we write.[11]
- **Books.** There are excellent books available to guide writers in their process, and we remind you of our annotated recommendations in the appendix. In the current context, though, we'll reinforce the idea that many developing writers are unaware that these resources exist, or are

reluctant to take advantage. In particular, we've heard grad students (and our colleagues) argue that they don't have time to read books on writing because they're too busy writing. But such books really can help with a writer's long game.

We could have made this list a lot longer, but it would be dull. The more important issue is what a mentor might do with such a list. Passing it on to the developing writers you mentor is a start, but only a modest one. We think the key is to see that they use the tools and do so with deliberate thought about which tools work for them, when, and why. In a classroom setting, this might be nicely accomplished by assigning students to use a tool, or several, and then to write a reflection.[12] What writing challenge can the tool help them address? Did it work well? What were its shortcomings? Can they think of other solutions to the challenge? For mentoring outside the classroom, we suggest a group meeting focusing on writing tools (for example, a dedicated session of the weekly lab meeting many research groups hold). We've found it particularly useful to encourage students to share their writing challenges and experience with tools they've used to address those challenges. This helps because there's a nearly universal pattern: students share many of the same challenges but have experimented with different solutions to them.

Text tools

Text tools—those that help shape the style and/or content of the text a writer is producing—are often considered quite different from process tools, in a pedagogical context, because they can be used (misused!) to avoid improvement in writing ability instead of to foster it. We encourage you, though, to take some historical perspective on this. Bethann and Steve are both old enough to remember the angst that seized both educators and parents when electronic calculators became cheap and small enough for students to carry to their K–12 math classes. There was a widespread opinion that mathematics education was doomed and that an innumerate populace was the inevitable consequence. It's true that on-paper long division has become a rare skill, but we still put rovers on Mars.[13]

There was nothing unique about the introduction of the calculator to education. New tools have been elbowing their way into the classroom for centuries: the textbook, the blackboard, the slide rule, the filmstrip, the microcomputer, PowerPoint.[14] Most new tools have been received with suspicion, most have then been integrated into pedagogical practice, and most are now used by teachers and students without a second thought. If you've been following the last few years' breathless discussion of ChatGPT and other "artificial intelligence" writing tools (see below), this perspective is worth remembering. It always comes back to *how* a tool is used, and many tools can be used to both improve the skills of developing writers and ease the burden on their mentors. With this in mind, consider some of the text tools that are available—some so familiar we don't even think about them, some so new they make headlines.

- **Spelling and grammar checkers and self-editing tools.** Nearly any modern writing software—even the texting app on a cell phone—includes built-in spelling- and grammar-checking functions. These have become better over time, but still make frequent errors of both omission (failing to flag mistakes) and commission (flagging as a mistake something that's completely acceptable). Developing writers should therefore be advised to think of built-in checkers as complementing, not replacing, their own proofreading and their own judgment. For better (but not perfect) results, stand-alone software or purposely engineered word-processor plugins include Grammarly and Linguix, along with many others. There are also more specialized self-editing tools (such as the Hemingway app and Helen Sword's *Writer's Diet*), that focus on helping a writer reduce complexity.[15] Most of these tools provide free and paid versions; in our experience, the free versions provide sufficient support for most students, and thus we do recommend them while clarifying that we don't expect students to pay for them.
- **Readability analyzers.** There's a profusion of software tools that compute "readability scores" for texts (for example, Readable, Datayze, and Lexile[16]). Readability is distinct from grammatical correctness (although grammar checkers may have readability scoring built in), in that readability focuses on the age or grade level at which an average reader would be able to engage productively with the sample text. The Flesch Reading Ease and Flesch–Kincaid grade-level scores are the most familiar, but many others exist and may be more suitable for writing intended for different audiences.[17] Some (like Flesch–Kincaid) are based only on word and sentence length, while others include information like the frequency of unusual words or the grammatical complexity of sentences. For developing scientific writers, readability analyzers may be most useful for assessing the match of text to its intended audience. This is most obvious for writers targeting general audiences rather than writing journal papers for other scientists, but we also see students using read-

ability analyzers to help them *increase* the complexity of their text! The latter practice follows belief (often fed by feedback from past instructors) that they aren't producing text matching expectations for Standard Academic English. In this case, we suspect most students will need some guidance to make sure they're making their writing more sophisticated, not more impenetrable. It's important that you help students understand that readability analyzers aren't assessment tools—there's no "right" reading level, except the right level for the intended reader.[18]

- **Thesauruses and dictionaries.** Perhaps these seem too obvious to mention, but the use of a thesaurus can either improve or weaken writing—and we see the latter frequently. Most developing writers know how to access an online thesaurus, and many do so with enthusiasm. The problem is twofold. First, the synonyms offered by a thesaurus aren't, as a rule, equally substitutable; instead, words differ slightly in meaning, and those meanings may depend on context or on the discipline being written about. A fine example is the confusion engendered by a "significant difference," which may be synonymous with "large difference" in many contexts but isn't in statistics (and thus, not in scientific writing). Developing writers should therefore be advised to use a thesaurus only in conjunction with a dictionary, and to think about which of several alternative definitions (and connotations) of a word might be inferred by the target audience. Second, developing writers often use a thesaurus because they believe varying word choice will make their writing more interesting. Variety must, however, be balanced with the increased cognitive load that inconsistent word choice places on the reader. A writer who refers to a "large difference" once, then a "major difference" later, asks the reader to think about whether "major" is larger or smaller than "large" and how the distinction might matter—and neither question might be intended by the writer. In sum: developing writers seldom need encouragement to use a thesaurus; but they may need encouragement to use it less, or to use it more effectively in combination with dictionaries and other references.
- **Writing centers, tutors, and editors.** A wide variety of *human* writing assistance will be available to most developing writers. In a campus context, there are writing centers and tutoring services (see chapter 7); away from the university campus, there are professional editors of various sorts (and, sadly, even "professional" essay-writing services). Developing writers may need help understanding what kinds of assistance are available to them, and also what kind of assistance is permitted (or ethical) in a given situation.

For a writing mentor, the first step is understanding what kind of resources are available to recommend. If you're teaching on a university campus, what will the writing center do for the developing writers you deal with? Will they read and comment on student drafts, or only offer general advice on structure and style? Do they address only work assigned in class, or will they help with other writing such as graduate theses or résumés? Do they offer workshops, one-on-one assistance by appointment, or drop-in hours?

Next, decide what kind of assistance you'd like to encourage, and what you'll permit (keeping in mind that if you disallow a particular kind of assistance, you need to think about how you'll enforce that[19]). This will likely depend a lot on your mentoring goals and on the kind of developing writer you're dealing with. For example, the use of off-campus professional editing services might be ethically appropriate for a graduate student writing a scientific paper, and might or might not be so for a graduate student writing a thesis,[20] but is generally agreed to be inappropriate for an undergraduate writing a term paper or lab report. Once you've decided what's allowed and what isn't, you'll need to articulate that to students so that it doesn't seem arbitrary or punitive—which risks engendering resentment, disengagement, or a deliberate flouting of your rules. Finally: we note that editing is a common tool in myriad professional writing contexts. There's nothing morally suspect or intellectually compromising about a student (or any other writer) seeking out professional support to hone their writing. The only question is how that fits with your mentoring goals.[21]

- **"Artificial intelligence."** While ChatGPT and its relatives belong on this list, they've received enough attention that they deserve their own, longer section. So, we'll return to "artificial intelligence" (and explain the scare quotes) shortly.

While these text tools may be quite different in purpose from the process tools discussed in the last section, we think mentors and instructors can think about their use in a very similar way. Two questions are important. First, how can developing writers be guided to use these tools both ethically and productively? Second, how can they be guided to use these tools to improve not just the text they're currently writing, but also their writing skills? At the risk of repeating ourselves, we'll point again to the value of assigning the use of these tools coupled with reflections about their use. We also encourage discussion (in class, in lab meetings, or otherwise) to help developing writers navigate when these tools should be used and how they can be used to best advantage.

Large language models (or "AI" writing assistants)

WHAT ARE LARGE LANGUAGE MODELS?

As we were writing this book, a new writing tool burst onto the scene: ChatGPT. ChatGPT was the first widely known example of a class of tools that are best called "large language models" (or LLMs). There is a more general term, "generative AI," that also includes tools that work with images and other media as well as text, but LLMs are most relevant to writing. Many more LLMs have since popped up, like mushrooms after rain: Microsoft's Bing AI, Google's Gemini AI, Anthropic's Claude, DeepSeek's oh-so-creatively named DeepSeek, and more. The "AI" (for "artificial intelligence") designation is a common shorthand, but it's both confusing (because lots of very different software tools get called "AI") and misleading (because LLMs are not at all what you'd normally think of as "intelligent"; see below). While LLMs are just another kind of text tool, analogous in many ways to thesauruses and paid editorial services, we've broken them out into their own section because they're new and (at the time of writing) controversial. A warning: language-model technology is advancing rapidly. While the general principles of what we say here are likely to apply for a long time, some of the details may be dated by the time this book is in your hands.[22]

Anyone who proposes to use, or to teach with or about, LLMs needs to start with understanding how they work, and thus what they can do and what they can't. LLMs are statistical models that, to oversimplify a little, are very good at producing English text that sounds like a human wrote it. They do this (again, oversimplifying) by taking a prompt and "knowing" what words—in the huge corpus of real human writing they've previously digested—usually follow that prompt. Initially, the prompt is your input to the LLM; then, as it produces text, the prompt for each new word or phrase is the passage that precedes it. If this sounds rather like the predictive text feature on your cell phone's texting app or in Microsoft Word, that's because it is. Just much, much more powerful.[23] And just as your phone's predictive text has pitfalls because it doesn't *really* know what you want to say or what the words it's proposing mean, there's plenty of potential for LLMs to lead writers astray.

CAUTIONS ABOUT LLMS: FUNCTION, ETHICS, AND ATTITUDES

A developing writer who understands what LLMs are and how they do what they do should realize that there are several reasons to be cautious about using them. These cautions are central to any use of LLMs in mentoring writing. That is, good mentorship should help writers understand both *whether* to use LLMs and *how* to use them so that their use is ethical and effective. We see at least three kinds of considerations here, and we'll discuss them separately, although they're in fact interwoven to some extent: issues of LLM function, the ethics of using LLMs, and attitudes readers may have when they become aware that a writer has used LLMs.

Function. Once a developing writer understands that LLMs simply generate human-sounding text, based on patterns in other texts they've been trained on, the writer can be guided to two absolutely crucial insights. First, no matter how confident it sounds, an LLM doesn't "know" anything. Second, an LLM isn't responsible for what it "writes"—its user is. We'll expand on each of these in turn.

What we mean by LLMs not "knowing" anything is that they produce plausibly phrased and styled text, but they don't generally know or care whether that text is true.[24] As a result, LLMs can "hallucinate" ("invent" is a better term, but "hallucinate" has caught on) supposedly factual statements that aren't true. In academic writing, hallucinations include misstatements of all kinds, as well as literature citations and statistics that don't actually exist.[25] Because ChatGPT is so good at writing plausible text, these hallucinations may sound very convincing; for example, an invented citation might combine the names of scientists who have published together, supply a title related to what those scientists work on, and identify a journal in which they've published—but with the combination being completely fictitious. This verisimilitude means that it's not a trivial task for a user to detect and fix an LLM's hallucinations. Equally concerning, LLMs can—arguably, are designed to—reproduce the racial, gender, ideological, and other biases that are present in the materials they trained on. If, for instance, prior human writing tends to represent scientists as mostly White men, an LLM may take that generalization and amplify it, writing *only* about scientists as White men—and so on.[26]

When an LLM hallucinates, who is responsible? Developing writers—all writers, actually—need to understand that if they use an LLM, *they* are entirely responsible for any mistakes it makes. If a term paper or a manuscript includes an imaginary citation, the fact that it came from an LLM doesn't excuse the author's decision to include it (and we would argue that inattention to the LLM output isn't an excuse, because that inattention is itself a decision). What comes out of an LLM, or any other writing tool, is only a suggestion. A writer may consider taking that suggestion; but they must make a well-reasoned decision to take it and then stand behind that decision.

Ethics. Any tool comes with ethical considerations, and while these can be unclear for LLMs simply because they're new, we suggest applying the same ethical standards you'd consider for any other writing tool, and asking developing writers to apply them as well. The idea that writers are responsible for an LLM's output is a good place to start. If a writer misses errors, distortions, or biases that an LLM introduces, that increases the workload placed on mentors, coauthors, peer reviewers, and other readers. It's not the job of any of these people to notice and fix problems arising from LLM use. Indeed, it's poor practice at best and unethical at worst for a writer to put others in a situation where they might feel they need to backfill for a writer's reliance on an LLM.

There are broader ethical considerations around LLM use, too. Like any other tool, LLMs have environmental and social costs.[27] The environmental costs of LLMs are widely discussed (often with claims of apocalyptic impacts) but complex to assess. Because both training and running LLM models is computationally intensive, they have costs in energy consumption and water use. The global *total* energy consumption, and thus carbon footprint, of so-called AI in general is startlingly large, but here we focus on LLMs in particular. Their energy consumption *per query* is modest—three recent estimates suggest a range of 0.007–3 grams CO_2, depending on the particular LLM used and the query made,[28] which may be dramatically lower than the carbon footprint of a human executing the same writing task.[29] (For comparison, watching an hour of streamed video generates approximately 30–100 g CO_2; driving a passenger car 1 km generates around 250 g CO_2; and you don't want to know about cheeseburgers.) Similar comparisons can be made for water use.[30] Nonetheless, because it's easy and fast for users to make many queries, and because of the tendency for LLMs to become embedded in other situations (such as web searches), some people find LLM energy demand concerning. Social concerns include the fact that the models are trained on a corpus of human-produced work whose authors aren't asked, acknowledged, or compensated. It's not entirely clear, either legally or ethically, whether authors *should* be compensated for LLM training use; as we write, multiple lawsuits addressing this issue are working through the US court system.[31] (A parallel issue arises when you consider uploading some text to an LLM as part of the prompt you're giving it: If the text is copyrighted, is uploading it legally permitted?[32]

What will the company providing the LLM *do* with that uploaded text, and even if copyright isn't a concern, is that consistent with the wishes of the text's author? If the text is your own, is uploading it consistent with *your* wishes?) A serious concern that's less widely reported is the psychological toll on human workers hired to find and label hate speech, violent threats, and more in LLM training corpora, to reduce the occurrence of such undesirable material in their outputs.[33] A classroom or lab-group discussion of LLMs, or an assignment involving using one, could include explicit attention to these and related issues, ideally with students charged with researching and considering multiple positions on one or more of them.[34]

One final point about ethics: users of LLMs shouldn't misrepresent their work. Unfortunately, it's not entirely obvious what constitutes misrepresentation in this context. There are plenty of other writing tools that can be used without disclosure to readers: grammar checkers, thesauruses, clever and articulate friends.[35] But to many, LLMs are different, and so it's probably best at least for now to err on the side of transparency: developing writers should explain their LLM use to mentors, coauthors, and supervisors. Certainly, they should do so when policy requires it. Many courses will require disclosure if an assignment was written with LLM assistance (if they don't prohibit LLM use altogether). Many journals do, too.[36]

Attitudes. From our discussion so far, you'll probably gather that many people have strong reservations about the use of LLMs. Actually, that's an understatement. Developing writers should be helped to understand that if they use LLMs in their writing, some readers (not all, and we lack data to quantify "some") will think poorly of them for that decision. Some may even refuse to read or cite work that acknowledges LLM assistance. (At least, some people are claiming to have such a policy, mostly on social media; of course, it's worth keeping in mind that social media favors nuance-free opinions.) A lot of these attitudes seem to arise from a belief that LLMs will be used to *replace* the human writer rather than to *assist* the human writer. It's the latter possibility we discuss in this chapter, but you should be aware that plenty of folks seem to resist recognizing a difference. So, our final caution is this: part of the job of mentoring developing writers is to help them understand reader attitudes about LLMs. Those attitudes will include very strong stances both in favor and against, and the mix is especially likely to be unsettled while LLMs are still new.

Teaching and mentoring with LLMs

Keeping in mind how LLMs work and the ethical considerations their use invokes, how might you use them in mentoring writing? How might you guide developing writers in using them in ways that don't just allow them to duck hard work? We don't recommend avoiding the topic, no matter how strong your personal reservations about the technology might be. Simply put, students *will* use LLMs. Even if you prohibit their use in your course, students will use them in other courses and later in their careers, so it's best if they're equipped to use them effectively and ethically.[37] (If you're tempted by prohibition, you should consider how you'll detect LLM use, how you'll navigate the issue of inevitable false positives, how you'll penalize LLM use, and how all this can be accomplished without consuming vast amounts of your time and energy.[38])

We think the crux of the issue is this: if you're going to use (or guide students to use) LLMs, you should be using those tools to build student writing skills—not to help students avoid building those skills.[39] But how?

We'll start with some general observations. First (to revisit a topic), since LLMs shouldn't be trusted to generate content, students should be steered away from asking them to review literature or critically evaluate an idea—unless asking students to evaluate an LLM's response is part of the assignment. Instead, students should be guided to think of LLMs as idea generators and style assistants. Second, one major strength of LLMs is that they can quickly generate multiple alternative versions of a text. Harnessing that capacity can allow students to consider strengths and weaknesses of alternatives and (more subtly) guide them away from the misconception that there's only one perfect way to write any given piece of text. Third, an LLM can be directed to answer queries in a particular way or with the adoption of a particular persona. For example: you can ask the LLM to simulate a kindly postsecondary writing instructor or a harsh reviewer—and these will suit different mentoring goals for different sets of students.[40] You can ask an LLM to provide edits, make suggestions, or ask questions. You can ask one to explain why a particular suggestion or edit might be recommended (although it may often be better not to, forcing students instead to think for themselves about why they might adopt the suggestion or reject it). Fourth, students may be willing to criticize and correct drafts they know were generated by an LLM where they would feel uncomfortable doing the same with drafts written by classmates or instructors. Finally, students may be more willing to solicit feedback from LLMs than they are to seek feedback from a human mentor, as the latter can make them nervous about exposing their perceived inadequacy.[41] They can also get rapid and repeated feedback without feeling like (or being!) a burden.

Now it's time for more specific suggestions for using LLMs in your mentoring. We'll begin with some exercises that are most obviously applicable in the classroom. While we acknowledge that the application of LLMs to mentoring writing is too new for much evidence-based guidance to be available, these suggestions are compatible with what's known about pedagogy more generally (for instance, you'll notice lots of student reflection).[42] We sketch these suggestions only briefly, so you can build on them to develop exercises that will work in your own teaching or mentoring context. All these suggestions are intended to help students use LLMs to think about writing, rather than to use LLMs to avoid thinking about writing.[43]

- Ask students to use an LLM to generate two alternative outlines for an essay or paper, given some bullet points about its content. Then ask them (the students, not the LLM!) to compare the two outlines, identify strengths and weaknesses of each, and then produce a third version of their own that builds on that comparison.
- Ask students to use an LLM to generate a draft of a passage (or full essay), including citations, and then have them fact-check the LLM's draft (focusing on content, not writing). To make this effective, you don't want the LLM to do too good a job; it will help to use a topic that's controversial or that involves very recent events. Alternatively, you could generate the draft yourself, asking the LLM to include factual errors. Ask students to consider accuracy from four angles. Is the LLM's draft reasonable overall? Does it contain specific instances of hallucination, including factual errors or the citation of nonexistent sources? Would they consider the draft to reflect a knowledge base that's deep or shallow (and up-to-date or not)? Does the draft show signs of embedded bias?
- Ask students to use an LLM to generate two or three drafts of a passage (or full essay), and then have them revise the texts into a single one of their own, analyzing for consistent, effective writing style, vocabulary, and structure. You can guide the students to a prompt that will likely engender writing issues: for example, asking an LLM to write "like a formal scientific paper" will probably produce the jargon, passive voice, and other infelicities the LLM experienced in its training corpus. You could also produce a problematic draft yourself: ask an LLM for an initial version, and then ask for a revision that's more formal and more technical, repeating a few times if necessary. Ask students to write a reflection

about what features of the LLM's writing they opted to keep and what they changed (and how). You could also ask them to reflect on the approach and workflow they took to tackle the revision.

- Ask students to submit a draft piece of their own writing to an LLM, asking it to revise to improve paragraph structure, organization and flow. (These are issues developing writers often struggle with). Ask the students to annotate the revision, indicating for each change whether or not they believe it improves structure, organization, and flow, and why.
- Ask students to submit a draft piece of their own writing to an LLM, asking it to revise using alternative word choices. Challenge students to use dictionaries to explain differences in meaning and connotations of their words versus the LLM's, and to defend their choice of one or the other. (This exercise gets at the tendency of some developing writers to use a thesaurus uncritically.)
- Ask students to submit a draft of their own writing (one that's fairly advanced in content and organization) to an LLM for copyediting of grammar, spelling, sentence structure, and other minor issues. Students should then generate a list of LLM-diagnosed errors that they recognize as ones they make often in their drafts, and will thus resolve to check *themselves* in their next drafts. They should be reassured that even the most experienced writer could make such a list (and many of us do).
- With permission, choose a real piece of student writing and then give an LLM the same assignment prompt that the student worked from. Have the LLM generate two or three versions of its own. Ask students to compare versions (including both theirs and the LLM's). Can they identify which is LLM-written? What are the strengths and weaknesses of each?
- Choose a piece of complex and technical writing, such as a published paper's abstract or an excerpt of a paper's Discussion, and have an LLM produce a plain-language version (perhaps at seventh- or ninth-grade reading level). (Choose the piece of writing you use so that it's legal where you work to use it as an LLM prompt.[44]) Ask students, before they've seen the LLM version, to do the same thing (note that many word processors, including Microsoft Word, can assign Flesch–Kincaid readability scores).[45] Ask students to compare, sentence by sentence, the choices they made against those the LLM made, discussing how those choices affected both readability and any loss of information or nuance in the plain-language version.
- Choose a piece of writing either from a less technical section of a scientific paper (the opening paragraph of an Introduction works well) or

from a work of popular nonfiction. Ask students to submit it to an LLM, asking for revisions in the writing voice/style of several well-known authors. Include authors of both fiction and nonfiction (for example, Ernest Hemingway, Toni Morrison, Bill Bryson, and Hope Jahren).[46] Ask students to write a reflection about what each style does differently and what audiences those features might be effective for. Ask them, further, to consider whether voice/style is evident in technical writing in their field (scientific papers, policy documents, etc.)—and to speculate on why it is or isn't. This exercise could set students up for an LLM-free extension to the lesson in which they compare popular and scientific writing by authors who write across genres (e.g., Hope Jahren, Chanda Prescod-Weinstein, Sean Carroll, or even either of us.) This extension is a bit more advanced, because students will be comparing material in which both content and style are different; the strength of using the LLM is that style can be varied for the same content.

While these exercises are designed primarily around written assignments by individual students, they could be adapted for a small-group workshop format, with student discussion replacing or complementing written reflections. Most could be adapted similarly for mentoring outside the classroom, perhaps involving a short series of lab-group meetings or similar sessions. In a lab group, with graduate students and other more advanced developing writers, you could build on these exercises by giving students more latitude to direct their exploration of LLMs. You could, for example, ask participants to experiment with an LLM, read some literature, and bring their best cases for and against it as a tool (considering both function and ethics). While this could be set up as a debate (with some participants arguing for and some against), it will be more effective to have each participant consider arguments for and against, then state a more nuanced position.[47]

However you envision student use of LLMs, you need to set up clear expectations. For courses in the classroom this might be a syllabus section, separate policy document, or special in-class discussion; for grad students and the like, a dedicated lab meeting is worthwhile. A course or lab policy might include a wide range of issues.[48] For what assignments or uses are LLMs permitted, and for what are they banned (and why)? Do you expect students to describe their LLM use in detail—for example, what prompts they used—or just disclose that they used it? Do you expect drafts or assignments to be submitted along with a reflection on how the writer used the LLM and what features of the output they found useful, rejected, or changed? If so, will these reflections be assessed (and how)? Perhaps all this sounds like extra work. It is. But in the long run, planning for your students' use of LLMs will *reduce* your need to police it while offering opportunities for better learning.

WHAT ABOUT WAYS THAT MENTORS CAN USE LLMS?

The most obvious possibility is asking an LLM to help with assessment and feedback.[49] While the prospect of this makes us deeply uneasy,[50] we see the potential and have colleagues who are already doing it. An LLM could be asked to comment on a student draft's organization, flow, complexity, and other writing features; it could even be fed a rubric along with a draft and asked to evaluate the work. Our most important caution here matches what we noted about the use of LLMs for writing: they will be very good at generating feedback that *sounds* like a mentor wrote it, but they won't care whether that feedback is accurate or helpful. Thus, you should always see an LLM as an idea generator; it can help you with what you *might* say but should never be used to give students feedback directly. We also recommend that if you intend to experiment with an LLM for feedback, you begin by hand-evaluating several texts before looking at the LLM's take on them. A comparison will tell you something about what LLM assessment is likely to miss (or what you're likely to!) and thus shape your use of the tool.[51] Beyond these functional considerations, there are obvious ethical issues in uploading student work to an LLM (just as there are with uploading student work to a plagiarism detector). Do you know whether students' work will be retained by the LLM, and if so, how it will be used? Are you (and your students) comfortable with for-profit use of the work? Do you have student permission to upload what they've written?

Something we hinted at in our list of LLM exercises for students is that you might use an LLM to generate text for students to work with. We've experimented with this and found it useful. With a bit of attention to composing the prompt, and perhaps with a few passes through the LLM, you can generate a passage on a topic that will be relevant to your students and that incorporates the kinds of writing strengths and weaknesses you'd like your students to diagnose, discuss, or fix. We've found that LLM assistance can make generating such passages much faster, and can produce passages with the kinds of errors that more accomplished writers find difficult to produce. However, as usual, you'll need to exercise some caution and supervise the LLM output.[52]

On another front, LLM-powered software has appeared recently that can generate lesson plans and similar documents from your learning objectives.[53] We haven't yet been impressed with their output, though they may be helpful as idea generators.

We'll close by echoing something we said early in this section: just as for student use of LLMs in writing, the application of LLMs to mentoring is so new that evidence for effective practice remains scarce.

New tools, new competencies

So far, we've discussed writing tools as just that—tools for writing. In other words, we've been considering the use of those tools simply as ways to improve writing and writing capabilities. But there's another perspective. At some point, a writing tool becomes widely enough applied that it becomes a competency that we might want to teach for its own sake.[54] Familiarity with project-management software, just as an example, may be something students should gain for more reasons than that it can help them with writing. Similarly, it seems clear that an understanding of what you can and can't expect from the output of large language models will soon be an important competency, whether or not one uses those tools for writing. We know this can seem discouraging; curricula are already stuffed with disciplinary knowledge, and the last thing many of us want is *more* new teaching and mentoring responsibilities. But there's nothing new about the press of new things. It's worth remembering that when we're mentoring writing, we're not just making our students better writers, but also better scientists and better citizens.

10: Writing in a Broader Curriculum

Can writing instruction be thought of as part of a curriculum rather than merely part of a particular course? • What might writing instruction look like if you could design it to run through an entire degree program or your lab-group training? And what kind of obstacles might you face in doing so? • How might you coordinate writing instruction so mentors and students aren't always dealing with conflicting advice? • Won't there still be coordination problems—only now, with advice from those outside the degree program or lab?

Students don't learn to write in a single course or from a single mentor. That's one reason the Writing across the Curriculum movement exists (WAC; see chapter 4). Ideally, writing instruction is integrated and coordinated throughout a degree program. In this chapter we take a direct look at how that might be done, in science curricula at both undergraduate and graduate levels. Our emphasis is on embedding writing assignments, both small and large, in content courses across a whole curriculum or through a lab group's activities. We discuss formal writing requirements, the availability of writing support and courses, and the complexities of mentorship by instructors and supervisors. We also address an issue of acute interest for most mentors and developing writers: the likelihood of contradictory expectations and advice between, say, two courses or two mentors. Students find this tension extraordinarily frustrating. And, while a range of advice is to be expected in learning to write scientifically, we propose ways to achieve course-wide, lab-wide, and department-wide consistency.

Writing is everywhere and nowhere

We'll start with a bit of a case study. Students begin Bethann's courses by stating their personal learning goals. The majority identify getting better at writing as a primary goal, even though most of Bethann's scicomm courses aren't writing courses per se. Many of these students are acutely—even debilitatingly—self-conscious of their limited progress as writers.

Sadly, for most of these students, it's arguably too late: by the time they enroll in Bethann's course, they're within one to three semesters of graduating. Most of their academic writing career is behind them, as is most of our opportunity to support their growth as competent and confident writers. The same can be true for graduate students, although perhaps to a lesser degree. For example, Steve's manuscript-writing course is an elective in which students often enroll late in their degrees, when the threat of a thesis looms.

As much as they dread writing, students at all levels know very well that they need to get better. Unfortunately, much of the experience they've had in high school and university was counterproductive (last-minute binge-writing an essay; never experiencing the value of revisions, etc.). They come to our courses hopeful, wanting (needing) to believe that one course can fill the gap. We try, and we're sure you do the same. But we know and you know the grim truth—it's simply not possible for a single course to transform a developing writer into a polished, professional one.

We're being blunt about this. Right now, students graduate unprepared for the writing responsibilities of their careers. And that's the fault of the departments and organizations training them. The good news is that it's possible to do better for developing writers by integrating writing instruction *throughout their entire learning journey.*

There's nothing revolutionary about this idea. Writing is such a foundational skill that developing writers receive assignments and advice from many different mentors, both sequentially (from kindergarten through graduate school and beyond) and simultaneously (in various courses, and from committee members, collaborators, and colleagues). What's often missing is deliberate coordination. The pathway from a child proud of their first words and sentences to adults experiencing procrastination, anxiety, and self-doubt is predictable and has real bearing on the problem.[1] Worse, the haphazard, course-by-course writing instruction we later provide doesn't build a transferable skill set that students can leverage from course to course and beyond academia. A lack of coordination also gives students the impression that writing isn't a core piece of the work of being a scientist. Instead of seeing writing as a central tool of scientific success, students come to perceive it as an obstacle to their professional progress.[2]

The story can and should be different. We owe it to our students (and the future we hope they shape) to embed coordinated writing instruction throughout our curricula. The ubiquity of writing in education is a huge opportunity, if we can shift it from a cause for hopelessness to a means of deliberate growth. We have some propositions for you about how to do that.

A warning: some of what we propose next may sound daunting. We encourage you to just do what you can; small steps to improvement are always better than no steps at all.

An ideal writing-embedded program structure (and how to shift toward it)

In table 10.1, we outline how we think a coordinated writing-embedded program could work for degrees that are course-intensive. This includes most undergraduate programs, course-based graduate programs, and some research-intensive graduate programs that also involve considerable coursework. Later in the chapter, we detail how you could manage a sequence to support students and others you mentor in your lab. This latter framework is important because you may not be able to implement a writing-embedded course sequence, yet your mentees still need consistent, sustained support.

To be clear: the course-based approach in table 10.1 is our dream program for coordinated, scaffolded writing instruction. It's based on our own experience and on a lot of reading of literature in pedagogy, writing studies, and rhetoric and composition, plus extensive input from the hive mind of social media. To our knowledge, this type of program does not actually exist anywhere,[3] so we can't point you to evidence directly assessing its efficacy. (It's common in K–12 education for curricular programs to be designed, piloted, and then rigorously evaluated for outcomes. The same steps are rare, at best, in higher education.) We'd love to see it put to a genuine test; in the meantime, we present it as something to embrace, tinker with, or even react against. Our hope is that considering this model at all will help you envision something that works for your students and your programs. After all, it's easier to imagine changes from a starting point than to conjure up a systemic approach from scratch.

In our model, we weave in multiple instances of the following themes:

- scaffolded instruction, in each course and between them (e.g., building skills incrementally; revisiting key skills and concepts throughout curriculum)
- a structural approach to writing (e.g., TEA (*t*opic/*e*vidence/*a*nalysis) paragraphs, outlines/reverse outlines, iteration to build up full texts)
- iteration and the social and emotional dimensions of writing (e.g., mindset, motivation, revision, feedback, peer support)
- the understanding that writing development has many phases and writing projects have many functions (e.g., progression from comprehension to conceptualization to communication and application)

TABLE 10.1

A possible approach to embedding writing in a four-year science degree program.

We provide options for undergraduate and graduate degrees that are course-intensive. Two-year graduate degrees will need substantial compression.

UNDERGRADUATE SCIENTIFIC WRITING PROGRAM	SCIENTIFIC WRITING PROGRAM FOR COURSE-INTENSIVE GRAD PROGRAMS
Year 1	
One writing advice book and one style guide (and ideally, a dictionary) are adopted and required for all students, to be used throughout the entire degree program. Instructors and TAs should be familiar with, reference, and adhere to the style guide for instruction and assessment. ***Introduction to Scientific Writing* course** in the first semester introduces (1) typical scientific and professional writing genres, (2) common and unique attributes of each, and (3) productive writing habits and tools. Depending on degree level, students write: ▸ professional emails and collaborator communications ▸ job applications ▸ reflections re writing experiences, goals, etc. ▸ drafts, revisions, and revision reflections focused on genres common in the students' discipline. ▸ outlines and reverse outlines for content-focused writing (see next point)	
Focus second semester on shared writing goals throughout content courses See next row of table for details. (While an additional writing course for undergraduate students in their second semester would be ideal, we recognize this is unlikely to be feasible for most programs. So, we refrain from suggesting one here.)	***Conceptualizing Research through Writing* grad course** in the second semester introduces the role of persuasive writing in scientific work (research proposals, funding applications, job applications, etc.). Includes exploration of peer writing group models. Students write: ▸ professional communications (e.g., permits, data requests) ▸ rough draft of master's/PhD proposal as a funding application ▸ draft of literature review making the case for their proposed work ▸ revision and revision reflections ▸ outlines and reverse outlines
Focus for year 1 in undergrad content courses: ▸ clarity of expression, comprehension of content (limited or no line editing, but do require practice with self-editing tools) ▸ Methods and Results sections (including conceptualizing study design via writing) ▸ stylistically diverse examples of scientific writing (e.g., papers from a wide range of journals, etc.)	**Focus for year 1 in grad content courses:** ▸ clarity of expression, comprehension of content (limited or no line editing, but do require practice with self-editing tools) ▸ reflections on how content is relevant to students' proposal ▸ professional correspondence and collaborator communications

<table>
<tr><th>UNDERGRADUATE SCIENTIFIC WRITING PROGRAM</th><th>SCIENTIFIC WRITING PROGRAM FOR COURSE-INTENSIVE GRAD PROGRAMS</th></tr>
<tr><td>▸ benefits of peer feedback
▸ revision and revision reflections
▸ outlines and reverse outlines</td><td>▸ stylistically diverse examples of scientific writing (e.g., papers from a wide range of journals, etc.)
▸ benefits of peer feedback
▸ revision and revision reflections
▸ outlines and reverse outlines</td></tr>
<tr><td colspan="2">Year 2</td></tr>
<tr><td colspan="2">Writing/communication course: Students enroll in at least one communication-focused elective course (in or beyond the department). Possibilities include writing-intensive literature, social sciences, or humanities course, or course in graphic design, communications and journalism, etc.</td></tr>
<tr><td>Focus for year 2 in undergrad content courses:
▸ semester 1: Introduction section, literature reviews, persuasive writing (e.g., articulating need for a research project)
▸ semester 2: conceptualizing research through writing; Discussion section of manuscripts
▸ undergraduate research experience queries/applications
▸ outlines and reverse outlines
▸ revision and revision reflections
▸ recognizing writing rules vs. genre expectations vs. individual preferences
▸ transparency of instructor effort, struggles, and processes in writing</td><td>Focus for year 2 in grad content courses:
▸ reflections on how content is relevant to student beyond their proposal
▸ How persuasion and argumentation function in published science (critique of methods, identification of assumptions and values, conceptual frameworks, caveats in discussions, etc.)
▸ Outlines and reverse outlines
▸ Revision and revision reflections
▸ Recognizing writing "rules" vs. genre expectations vs. individual preferences
▸ Developing a capacity for self-editing
▸ Transparency of faculty effort, struggles, and processes in writing</td></tr>
<tr><td colspan="2">Applied Principles of Science Communication course for undergrad and grad students
▸ key concepts of sharing science beyond academia (e.g., there's no monolithic "general public," dialogue/co-production vs. deficit model, calibrating messages and approaches to specific audiences)
▸ conceptualizing, planning, implementing, and assessing an applied scicomm project
▸ peer feedback and accountability groups</td></tr>
<tr><td colspan="2">Year 3</td></tr>
<tr><td>Focus for year 3 in undergrad content courses
▸ undergraduate research experience queries/applications
▸ developing self-editing capacity
▸ polishing writing (concept, organization, clarity, style, and line edits)</td><td>Focus for year 3 in grad content courses
▸ networking through writing: informational interviews, social media, etc.
▸ Writing as research (vs. "writing it up" at the end)
▸ enhancing self-editing capacity</td></tr>
</table>

(continued)

TABLE 10.1 (*continued*)

UNDERGRADUATE SCIENTIFIC WRITING PROGRAM	SCIENTIFIC WRITING PROGRAM FOR COURSE-INTENSIVE GRAD PROGRAMS
▸ revision and revision reflections	▸ polishing writing (concept, organization, clarity, style, and line edits) ▸ revision and revision reflections
Year 4	
Capstone writing/communication undergrad course ▸ major writing/communication project of student's choice ▸ has an applied, beyond-the-classroom component ▸ revisiting and refining approaches to professional correspondence and collaborator communications and job applications ▸ polishing writing (concept, organization, clarity, style, and line edits) ▸ revision and revision reflections ▸ reflection on growth as a writer during their degree **Focus for year 4 in undergrad content courses** ▸ refining self-editing capacity ▸ polishing writing (concept, organization, clarity, style, and line edits) ▸ revision and revision reflections ▸ peer feedback and accountability groups ▸ co-writing/group writing projects	**Capstone writing/communication grad course** ▸ submission of at least one chapter for publication ▸ completion of full draft of another chapter or grant proposal ▸ adapting work for an applied, beyond-the-classroom component ▸ polishing writing (concept, organization, clarity, style, and line edits) ▸ revision and revision reflections. ▸ reflection on growth as a writer during the degree **Focus for year 4 in grad mentoring contexts (coursework likely complete)** ▸ revisiting and refining approaches to professional correspondence and collaborator communications and job applications ▸ refining self-editing capacity ▸ polishing writing (concept, organization, clarity, style, and line edits). ▸ peer feedback and accountability groups ▸ co-writing/group writing projects

- a commitment to students not feeling alone as developing writers, and normalizing the work of growing as a writer throughout the degree program
- communicating the work that experienced writers do when they write
- making transparent the nuances of writing conventions in distinct genres and disciplines (e.g., proposals are handled differently in different labs, departments, and disciplines[4])
- recognizing the relevance of writing as a transferable skill for scientific, professional, and civic writing

We've discussed the rationale for this framework in detail throughout this book, so we'll be succinct here. We embedded the central tenets of a healthy mindset about writing and a productive approach to the act of writing. But we aren't wedded to every detail, so we hope you'll see ways to tinker, prune, graft, and improve. We understand that you probably can't institute a program like this overnight! But we hope that seeing this model can help you and your colleagues take small steps (like adopting a common style guide) that will still make a big difference.

COORDINATING INSTRUCTION TO REDUCE CONFLICT

Our proposed curriculum probably raises an obvious question: how on earth can we ensure that writing instruction is consistent across so many courses and instructors? First, we have to reckon with an elephant. Developing writers are constantly caught in the crossfire of situations like this: Bethann adheres to American English styling of commas on the inside of quotation marks. Thus, each time she reads one of Steve's drafts, she "corrects" Steve's Canadian placement of commas outside the quotation marks.[5] You can imagine the chaos if Steve were to "correct" them back, and so on ad infinitum—and pity the unfortunate students taking both our classes!

Of course, our comma example is trivial, but it represents a wider problem. Along with stylistic or grammatical tensions, students encounter much more consequential conflicts that may impact conceptualization, framing, analysis, rhetoric, and more. Such tension can happen in other spheres, of course (statistics!), but writing is particularly vulnerable because many instructors hold strong yet unsubstantiated opinions about "correct" writing. Students may wonder how they can possibly succeed in two different courses with conflicting expectations. Beyond this immediate concern, it may be difficult for students to learn which issues actually matter to their lifelong writing skills. Undergraduate and graduate students are embroiled in such dilemmas *all the time*, and the issues persist to become a dreaded aspect of coauthorship and peer review (see chapter 5 for our treatment of such conflicts outside the classroom).

The solution is simple in principle, although we admit that it's difficult in practice: instructors of multiple courses (ideally, an entire degree program) must agree *in advance* on their writing expectations. But how?

We begin with a possibility so obvious that we're surprised not to know of any department that has done it: the adoption of a common writing guide, style guide, and dictionary (or at least the writing guide) that students can turn to throughout their degree programs. We're so accustomed to a textbook being required for a course that perhaps it doesn't occur to us that we could require one for a degree program! (This, by the way, is how Steve and

Bethann resolved their comma dispute: by agreeing to follow the style guide of the University of Chicago Press.) There are good, affordable writing and style guides. While no department will ever agree in complete detail on the contents of any of them, that's no different from any other curricular issue, and a department can agree to follow one of them regardless.

A caveat: deciding on a single writing guide should dramatically reduce instructor-to-instructor conflicts, but it risks misleading students into thinking that one book contains the inviolable rules of writing. Also, your students will inevitably run into conflicts anyway; even if you remove them in your own degree program, you can't do so for the broader world. Instructors should therefore make clear that most aspects of writing are matters of genre conventions and style, rather than hard-set rules. Through assignments designed to highlight this point, students can be coached to anticipate that each mentor and each writing task will involve a blend of preferences and settled standards. Equipped with this awareness, students can avoid some frustration by knowing they'll need to understand expectations for each writing task—as they will throughout their careers.

More ambitiously, we can move beyond the simple adoption of a common writing guide toward a committed integration of writing instruction throughout an entire degree program.

That's why we've provided (in table 10.1) a full four-year model that you can explore. But we won't leave you with a supposed Utopia; we also consider some of the reasons why this kind of program might be challenging to implement, and why some of your colleagues are likely to resist. The effort of (re)designing a degree program with integrated writing instruction will undoubtedly cause all sorts of tensions, pet peeves, and unfounded notions to bubble up alongside the evidence base and expert guidance you may consult. The potential payoff, though, is enormous. Collectively agreed-upon writing instruction reduces conflict and workload for the developing writer *and for those who mentor them*—all while enhancing learning. That's a win-win.

SOME CHALLENGES YOU MIGHT FACE WITH AN EMBEDDED-WRITING APPROACH

Academics are, in our experience, pretty darned good at coming up with reasons something can't be done. So we know that if you propose something like the program above, you'll get pushback. Here we anticipate five particularly likely kinds of resistance and suggest some ways you can respond.

1. It may be hard to fit writing into content courses without more specific guidelines.

We realize we haven't given you a complete how-to. You can do a better job than we can of harnessing the central principles of our framework to design detailed course syllabi for your particular programs and your particular students.[6] That's especially true for content courses (versus writing-focused ones), where there will be major differences across disciplines. If you want focused prescriptions for your program, you can refer to chapter 7 for suggestions about resources you can draw on, including other instructors' syllabi and colleagues on your campus.

2. It may be hard to get students to take all these courses.

The model we suggest recommends writing instruction in content courses, writing-focused courses taught in the degree program (three for undergraduates, four for graduate students), and at least one external, communication-focused elective course. We recognize that each degree program and university will have constraints on courses that can be added to degrees. These include:

- Some programs may already have so many content courses that it would take a major overhaul to add writing-focused ones.[7] However, others may encourage external electives to facilitate engagement with other disciplines.
- There may also be constraints posed by accreditation for degrees aligned with external, professional certifications.
- Students transferring in from other institutions may land halfway into your curricular progression, and you'll have no control over their previous writing training.
- Some graduate programs require very few courses at all. This is particularly true at the PhD level; indeed, in some research-focused graduate programs, students may take only a couple of credit-bearing courses during their entire program. For these situations, you might consider using discussion groups, writing groups, and the like to provide the conceptual and practical support we recommend. (See table 10.2 for some models.)

We don't offer detailed prescriptions of how to address all these enrollment issues, because your situation is unique to your institution, program, and student body. We encourage you to work to address them in ways that provide consistent writing instruction for your students.

3. Assigning teaching for the writing-focused courses is a challenge.

Finding instructors willing and able to teach writing-focused courses can be a concern. It's not just a matter of finding qualified instructors; the faculty who teach in a program may already be overwhelmed by offering the discipline's content courses, and university administrators usually want to cut faculty positions rather than bolster the ranks. Even so, there are at least two possibilities. One is that administrators can be persuaded to support the kind of innovative program-wide revision we're suggesting (with resources including faculty lines) precisely because it *is* innovative. Another is for science programs and departments to leverage a model common in writing programs and English departments. In those units, introductory writing courses are typically offered in multiple sections, with each taught by a graduate student who is the official instructor of record.

The multiple-grad-led-sections model is possible because graduate-student instructors participate in a training program that orients them to the curriculum before the fall semester. Then, during their first semester of teaching, they are simultaneously enrolled in a writing pedagogy course (three credits). Each semester they teach, they (ideally) also participate in a writing-instructor mentoring course (one credit). This doesn't reduce the instructional load for faculty to zero, but it spreads out the work and lets faculty focus on training and overall coordination. Thus, it addresses a major constraint most programs face: more undergraduate students than can possibly be taught in the hands-on, small-class-size approach recommended by most writing research.[8] Ideally, this model would be built out in consultation (or collaboration) with pedagogical and writing-instruction experts from your campus.[9]

We don't want to leave you with the impression that we're just leaning on graduate students to do the work. They benefit, too, because they get valuable teaching experience beyond TAing, and receive instruction in general and writing-specific pedagogy that's otherwise rare, plus sustained mentorship throughout their teaching. In particular, the mentorship course supports grad students to deliver consistent feedback and assessment across sections, helps them navigate the inevitable teaching challenges all new instructors experience, and builds a sense of community among student instructors. Furthermore, teaching-to-learn is a time-honored approach to improving one's own skills, and teaching writing is a powerful way to become a better writer.

There are, of course, potential obstacles. For one, faculty unions (where they exist) may balk at courses taught by those other than union members. For another, graduate programs with very little coursework may not easily accommodate the amount of training it takes to ensure that the graduate students are effective instructors. It's also just plain unfamiliar to most of us. (Bethann notes that she has proposed this model—in which she was trained—to her department for several years, and her pitch has not yet succeeded.) We've talked about it at length anyway, in part because it *is* unfamiliar, and in part because it offers large potential payoffs.

4. It may be hard to get everyone to use the same style guide.

Someone's going to say it, so we might as well. Disciplinary variation and territoriality may be significant obstacles to getting faculty within a program to use a common writing guide.[10] Getting your colleagues to this point might be a long game, but you needn't wait for that to play out. You can start on your own, if you instruct more than one course. Or perhaps you and a few like-minded instructors (ideally including those teaching first-year courses) do it from the grassroots: you agree on a guide and then gradually work to get more colleagues on board. You can also work on a core curriculum where students follow a set progression of courses within a broader program. Over time, the utility of the common style guide will become apparent, and that will help your longer-term effort to get an entire program on board. Finally, if you're a department, program, or curriculum-committee chair, you can decide the benefit to students is worth burning some social capital and push for change department-wide.

5. Colleagues may argue that scaffolding focuses too narrowly, that students need to write whole documents.

Our recommendations for scaffolding strike some people as leaving earlier courses incomplete.[11] What use is it, they might argue, to teach first-year students only about Methods and Results? Our reply: as we've discussed throughout this book, learning to write isn't a one-semester process. Scaffolding across courses recognizes this and aims for sustained growth in developing writers. Furthermore, breaking down competency in writing a lab report or research paper across courses and years doesn't preclude discussing the entire structure in a given course. It simply gives you and developing writers you mentor the permission to (1) recognize that the Methods section or Discussion is a part of a larger whole while (2) concentrating on a small-enough section of a larger writing project to allow time for iteration, feedback and revision, and metacognitive reflection. Another way to think about it: each section of a research paper is actually its own mini genre with specific conventions and norms. Focusing on one section (while in conversation with the whole paper) makes these nuances more explicit for developing writers.

The benefit of scaffolding isn't felt only by students, though. Many instructors don't feel confident placing a major emphasis on writing instruction. In a program where writing is scaffolded, each year or course emphasizes particular types of writing or habits of mind. For an instructor, focusing on specific aspects of writing instruction can make it more manageable.

There's yet another angle on the separate-sections-in-separate-years approach. One of the most common frameworks for gauging efficacy of a course or program involves defining learning objectives and then mapping out how content and skills will be introduced (I), reinforced (R), and facilitated toward students achieving mastery (M) in that area.[12] Typically, a single course will have learning objectives spanning the range of IRM, with some aspects emphasized in greater depth (toward M). The same progression is common in a given program's curricular map, where several courses in a suite are responsible for an IRM progression through program-level learning objectives. (If you're unfamiliar with mapping and assessing learning objectives, see chapter 4.) Whether they use the lingo or not, most departments and programs already have experience conceptualizing progressive instruction through a curriculum—and that's what scaffolding writing development across courses does.

A writing "curriculum" for mentoring individuals and lab groups

So far, we've considered embedding writing in a course-based curriculum at the level of a degree program. It's also profitable to think similarly about the mentoring you do with graduate students and other developing writers in your lab (for simplicity, we'll talk about grad students because they're the most common case). In fact, this approach may be more attainable for many mentors: it's a big reach to change your whole degree program, but relatively easy to set in motion a curriculum-style approach with an individual student or a lab group.

Just as we suggest for the course-based sequence above (table 10.1), you should begin by adopting one writing-advice book and one style guide for your lab (see the appendix for possibilities). These should be required reading for all students and trainees joining your research group, in the first three-ish months they work with you. (We know of PIs who buy multiple copies of a book and give one to each new student who joins the lab. They reason that this small investment is more than recouped via more efficient mentoring.) As the mentor, you should be equally familiar with, and largely adhere to, the guidebooks for instruction, feedback on writing drafts, and so on.

With this foundation, we suggest an explicitly designed multiyear "curriculum" (table 10.2) that you could implement either for a single student (left column) or for a group of students, such as your lab or a couple of like-minded labs (right column, plus activities chosen from the left column organized into lab-group meetings). We've laid it out as a two-year program, because that can accommodate both master's and PhD students, and we've front-loaded it because you'd like to see payoff as soon as possible (right?). You can, of course, stretch or compress as needed. If you're especially ambitious, you could even incorporate additional content from the course-based curriculum (table 10.1). If you're adapting both tables for your lab, we'd suggest rereading two earlier chapters as you do it: chapter 5 for approaches to mentoring writers outside the classroom, and chapter 6 for ideas about mentoring students toward learning on their own.

We recognize that this proposed "curriculum" may seem like a lot of mentoring work, but we suggest you think of it as systematizing and likely reducing work you're already doing. You're especially likely to find it paying off when a student reaches the stage of finalizing their research proposal or publishing their research chapters. The process we've outlined also gives both you and your students a clear road map so everyone knows what to expect regarding developmental milestones for a writer in your lab group. We close, however, by admitting that this (like our course-based curriculum) is utopic: many mentors do parts of it, but we don't know anyone who has (yet) implemented this kind of writing program in full.

What about conflicting advice from outside the program (or lab group)?

Beyond the logistics of your own program or lab group, there's a boundary problem—there always is. No matter how well you coordinate writing instruction in a lab, a set of courses, a program, or even a whole university, students will always hear advice from outside: from friends, family members, other mentors, readers of undergraduate or graduate theses, university writing centers, collaborators, programming from professional societies or at conferences, blogs/newsletters, and much more (see more discussion along these lines in chapter 7). Developing writers may benefit from

TABLE 10.2

A possible approach to sustained mentoring of writing for an individual or a lab/research group

INDIVIDUAL STUDENT	LAB GROUP *Adapt "individual student" activities as needed to round out the recommendations below.*
Year 1, semester 1	
Adopt a book of writing advice and a style guide. Ask the student to read the two books you've selected (ideally as a standard set for everyone in your lab) and to write a reflection about how these compare to their previous training and experience. Have them flag new terminology and anything else from the books that they have questions or concerns about. Discuss this reflection with them. **Program-specific writing requirements.** Many programs have set deadlines for things like research proposals or written comprehensive exams. Task the student with preparing a list of such writing requirements, along with target dates for completion of intermediate steps (first drafts, specific sections, etc.).[a] The student should build in scheduled discussions with you and can be asked to suggest when these would be most helpful. (This can also extend to student presentations for proposals, conferences, etc.) **Mindset development.** Use readings and discussions to introduce students to healthy mindsets about becoming a scientific writer. There are many topics you might include;[b] think about the needs and background of the student you're mentoring. You can find readings to support most of these topics in a writing advice book—so much the better if it's the one you've set as a lab standard. (We mention particularly recommended resources in the notes to this table.) Some possible topics: ▸ writing as a skill/craft to develop[c] ▸ your own experiences learning to write ("you" the mentor) ▸ your writing process (including drafts, rejections, coauthor feedback, and examples of writing in other genres like a blog, if you're active in that way)[d] ▸ social, behavioral, and psychological aspects of writing[e] ▸ supervisor/student relationships regarding writing[f] ▸ campus resources to support research and writing[g]	**Near-peer pairs (student-led).** Pair up each new student with a senior member of the lab as a writing mentor and with a closer peer as a writing buddy. Assigning each trainee to two people in the lab helps ensure they don't wallow in their own fears, isolated from the social dimensions of getting better at writing. It also helps spread out the burden of mentoring developing writers.[l] The pairs should check in at least weekly with each other to update, troubleshoot, review drafts, etc.; you can meet with them less frequently, perhaps monthly). Pairs should emphasize drafting and iteration, not achieving perfect, polished documents (at least not early on). This focus helps students get familiar and comfortable with the extensive revision required in academic writing. (See chapter 7 for some discussion of peer and near-peer review.) **Craft workshop (faculty-led).** Facilitate a weekly mix of discussing papers, books, and documents in other target genres (e.g., grants, job applications, etc.). The key is ensuring a focus on the craft of writing rather than only on content and avoiding "shark tank"–style teardowns.[m] For example, discuss how agreement, disagreement, and novelty are conveyed in the text; or discuss aspects of the writing that students would avoid or choose as a model. The craft approach also helps students practice the vital habit of omnivorous reading needed to become a better writer. If you have a lab member with enough experience to lead this group, all the better; this will help keep it fresh and productive for them and ease the load on you. You can build into the mix some sessions that don't depend on readings, too; for example, this would be a great venue for a discussion of writing challenges and the tools your students have discovered to address them (see chapter 9). You can also draw from topics in the left column.

INDIVIDUAL STUDENT

LAB GROUP

Adapt "individual student" activities as needed to round out the recommendations below.

- reading (both primary papers and literature reviews) and managing the literature[h]
- tactics for writing-group formation[i]
- critical reading as a form of writing and finding your authorial voice[j]
- key sentences and other exercises[k]

Students find discussion of these issues most helpful when it involves other students. If you can't arrange that for students you mentor, encourage them to find interested peers.

Writing group (student-led). Task your lab with implementing a weekly writing session. You can decide whether you want to leave the format up to them or be more prescriptive.[n] Be clear that the writing group doesn't replace their independent writing but will provide helpful structure and writing accountability.

Year 1, semester 2

Building blocks and foundational genres for scientific writing. Continue the sequence of readings and discussions from semester 1, ideally supplemented with appropriate writing exercises.[o] Full-scale writing assignments are likely too much for students already adjusting to the pace of graduate life, but consider taking advantage of program-specific writing requirements as they fit the topics to be covered.

Some possible topics:

- academic style and conventions in your discipline[p]
- major structural features and storytelling arcs in scientific writing[q]
- stylistic and structural variation (data papers from a range of journals, data papers vs. reviews vs. editorials, papers vs. scicomm)[r]
- approaches to revision (after self- and external feedback)[s]
- research project proposal conventions and expectations (this will be a priority in graduate programs putting heavy weight on the project proposal)[t]
- funding proposals, along with scholarship and fellowship applications[u]
- outlines and reverse outlines[v]

Near-peer pairs (continued). Ask students to write a short reflection on what they've gained from their near-peer experience, then share that reflection in a group meeting. (Writing it out first encourages more careful reflection.) Maintain near-peer relationships (troubleshooting if needed). And now also task the near-peer pairs with articulating goals and a development plan for the semester. At minimum, each pair should articulate two or three things they want to focus on. They can use books and readings from the previous semester to inform their plan, which should include a means of assessing their progress. (Bonus points if they make the development plan evidence-based, with connections to literature and tools that support writing development.)

Writing groups and craft workshop (continued). Ask students to write a short reflection on what they've gained from their writing-group and craft-workshop experiences. Maintain both activities with continued writing, reading, and discussions (sourcing activities from the left column and elsewhere in this book).

(*continued*)

TABLE 10.2 (*continued*)

INDIVIDUAL STUDENT	LAB GROUP *Adapt "individual student" activities as needed to round out the recommendations below.*
Year 2*	
Writing development plan. Each year, following a lab member's first year with you, ask them to articulate writing goals and a writing development plan (this can be part of their annual development plan, if that's something you do with them). Encourage the student to identify multimodal tools and resources to support their growth, including participating in workshops available on or beyond your campus, reading additional writing guides, pursuing opportunities to write in an array of genres, etc. **Further topics in scientific writing.** Continue the sequence of readings and discussion from year 1, with more advanced themes. Here, it will be useful to ask students what they wonder about (or struggle with), and to consider your students' intended career paths and thus what topics might be most important to them. Topics might include: ▸ polishing and presenting a research project proposal[w] ▸ practices and ethics of authorship[x] ▸ norms in citation practices, including ethics of citation[y] ▸ effective practices for collaborative writing[z] ▸ writing responses to reviewers[aa] ▸ professional writing genres beyond academia (e.g., reports, technical writing, etc.)[bb] ▸ cover letters for job applications and paper submissions[cc] ▸ scicomm (public-facing) writing[dd] **Co-review of a manuscript.** Any student envisioning an academic career (and quite likely others) will eventually be invited to serve as a peer reviewer. *With permission of the handling editor*, ask a student to co-review a manuscript you're already reviewing. Task the student with constructing a respectful, logical, and convincing review,[ee] which you then edit (combining with your own review) and return to the journal's handling editor.	**Near-peer pairs (continued).** Consider reshuffling near-peer pairs, especially as lab members who have gained experience become able to provide mentorship to new, more junior members. Maintain the expectation that these pairs meet weekly and check in with you monthly—it isn't just "new" writers who need regular feedback and support. **Writing groups and craft workshops (continued).** Ask your students to consider again their writing-group and craft-workshop experiences. What has worked well? What hasn't? Task them with implementing at least one change to their groups and writing an explanation of why they expect that change to be productive. (Bonus points if they make the rationale evidence-based, with connections to literature and tools that support writing development.) **Collaborative writing project.** Ask the lab group to work together on a paper, scicomm activity, or other writing project. A shorter project first may help the lab find a workflow and collaborative style that's effective—such projects might include a lab vision statement, text for the lab's website, or a job ad for a lab tech. Ideally, this isn't busywork, but a piece that will actually be published or otherwise used, thereby contributing to the professional portfolio, and possibly even thesis, of each student.[gg]

INDIVIDUAL STUDENT	**LAB GROUP** *Adapt "individual student" activities as needed to round out the recommendations below.*
Response to review. Task the student with writing a formal "response to reviews" in connection with a piece of their work that has received comments.[ff] They could work with reviewer comments on a manuscript, supervisory committee comments on a chapter draft, or even comments from a near-peer pair on a writing project of any kind. Afterward, you and the student should discuss their responses, including how they conveyed their resistance to suggested edits respectfully but robustly and how they considered whether disagreeing might put them at more professional risk than is worth it.	

* *We haven't split year 2 into semesters the way we did with year 1. That's deliberate; we think there's enough variation among programs and among students that you'll be the best judge of how to package this more advanced content for your students.*

a. Students often find it challenging to reduce a major output (e.g., proposal) into intermediate steps. You can refer them to Bethann's sequence for making this level of a to-do list (Merkle 2023b) or point them to apps that offer suggested subtasks for any item you provide as a prompt (e.g., Goblin Tools' MagicToDo app: https://goblin.tools/).

b. Fisher et al. 2024.

c. See chapter 1 for discussion of a growth mindset, as well as Fisher et al. 2024 for examples of how to facilitate such conversation. Heard 2022c, chap. 2, provides a short and accessible discussion of writing as a craft; King 2020 is a much longer reflection on the same topic.

d. For more discussion of how to share your own work, processes, and experiences with students in productive ways, see chapter 1.

e. Sword's BASE activity (https://writersdiet.com/base/base/) is a great starting point. Heard 2022c, chap. 4–6, provides a discussion of writer behavior and its intersection with human psychology; Silvia 2018 recommends scheduling and other interventions to overcome some common behavioral obstacles.

f. See chapters 5 and 6.

g. See chapter 7.

h. Heard 2022c, chap. 28, distinguishes between "reference," "survey," and "deep reading," and discusses techniques for each. Managing literature (including the use of reference-management software) is an area of strength for librarians; your libraries are likely to provide useful materials, or—even better—librarians can directly help mentor students (see chapter 7).

i. For useful discussions of writing group formats, see Rockquemore 2010 and Silvia 2018, chap. 4; for a précis, see our chapter 7.

j. See discussion in Heard 2022c, chap. 3, for a brief discussion of reading to write. Authorial voice is rarely discussed in scientific writing (but see Heard 2019a); Robbins 2016 and Rankin 2022 comment on voice in academic writing more generally.

k. Graff and Birkenstein 2021 is a valuable resource for students learning how typical moves in scholarly writing (e.g., agreement, disagreement, assertion of novelty) are conveyed. The book is written for undergraduates, but in Bethann's experience, graduate students also appreciate these models. Swales 1990 and 2004 provide deeper and more academic dives.

l. You'll want to ensure that these folks have some coaching on how to *be* mentors, and you'll either need to release them from other duties or explain how supporting a mentee is a lab duty and something that furthers their own development. (See more about this in chapter 7.)

m. Fisher 2024 and Mewburn 2013 (especially the Thesis Whisperer Verb Cheat Sheet) provide useful checklist-style documents that can draw participants' attention beyond what a text says, to focus on how it's crafted to say it.

(*continued*)

TABLE 10.2 (*continued*)

n. For more on writing groups, see chapter 7; For specific models for how writing groups can work, see Rockquemore 2010 and Silvia 2018, chap. 4.
o. There are many suitable exercises in Heard 2022c, Bean and Melzer 2021, and other books listed in the appendix. See also Fisher 2024 for a worksheet that helps students "read like a writer."
p. For a starting point, see Heard 2022c; his chapters 8–16 break down structure and style norms for IMRaD and other paper forms in the sciences. For resources you can use to talk about a broader range of genres, see the appendix.
q. See previous footnote. Schimel 2011 is also strong for storytelling in scientific writing. Dicks 2018 explores storytelling more generally.
r. Heard 2022c, chap 16, covers journal paper structures beyond IMRaD; chap. 26 covers, rather briefly, non–journal paper genres. For deeper treatment of non–journal paper genres, see resources in the appendix.
s. Heard 2022c covers self-revision (chap. 21) and revision following friendly (chap. 22) and formal (chap. 24) review. Belcher 2019 uses a more task-detailed, workbook-like approach to guide writers through structural (chapter "Week 9") and sentence-scale (chapter "Week 11") self-revision, as well as revision after formal review.
t. Because every program has its own requirements and norms for project proposals, resources we'd recommend might not fit well. However, see note u for resources covering funding proposals, which have obvious relevance.
u. Among scientific-writing guidebooks, Schimel 2011 and Hofmann 2022 cover funding proposals in some depth. Full book-length treatments include Oruç 2011 and Friedland et al. 2018. Heard 2022a is an entry point to a five-post blog series that might provide a short and accessible way to spark discussion.
v. Heard 2022c, chap. 7, provides detailed discussion about approaches to outlining; there are more ways to outline than most developing writers will suspect. For reverse outlining, see our discussion in chapter 3, along with this resource from the Brandeis University Writing Program: https://www.brandeis.edu/writing-program/resources/faculty/handouts/reverse-outlining.html.
w. See chapter 7, note 1.
x. Heard 2022c, chap. 27, covers the history and modern practice of coauthorship and discusses common criteria for authorship. Cooke et al. 2021 offers advice on avoiding authorship conflicts. Duffy 2017 explores authorship order in ecology; this can be a jumping-off point for discussion of authorship order in any discipline. Liboiron et al. 2017 looks at authorship order through an equity lens. For a particular focus on authorship in thesis chapters, given collaboration of student with supervisor and student with student, see Heard 2018 and Thomson 2017.
y. Heard 2022c, chap. 15, provides a brief summary of the nuts and bolts of citation in scientific writing. For more socially embedded discussion, begin with Mott and Cockayne 2017. Vettese 2019 provides a perspective on gender and self-citation that invites consideration of ethical remedies.
z. Virtually every writing mentor (that's you!), and many developing writers, too, will have deep experience with collaborative writing, and there's an enormous range of variation in practice. Heard 2022c, chap. 27, could be the starting point for discussion, but we would encourage you and your students to talk about what has worked and what hasn't, for you and for them.
aa. There are few published treatments of this, which surprises us because it's such an important part of the process of writing scientific papers for publication. Heard 2022c, chap. 24, provides a fairly detailed guide.
bb. See chapter 2 for detailed discussion about how to engage students in writing experiences relevant to their careers beyond academia, and chapter 4 for suggestions about connecting students to professionals working beyond academia and/or in nonacademic genres. See the appendix for some resources covering genres beyond journal papers.
cc. Hofmann 2022, chap. 31, covers writing for job applications, although we stress that expectations vary substantially among disciplines and are changing through time. Thus, you might complement written advice like Angie Hofmann's by involving a colleague who has recently served on a search committee in your discipline. There's lots of debate about how useful cover letters are for journal submissions, but most journals request them. Some journals have very specific requirements for their content; for example, they may require conflict-of-interest or funding disclosures or peer-reviewer suggestions. Taylor and Francis provides a detailed guide that could help you structure discussion; see their "How to Write a Cover Letter for Journal Submission," https://authorservices.taylorandfrancis.com/publishing-your-research/making-your-submission/writing-a-journal-article-cover-letter/.
dd. See especially Nijhuis 2016 and Kearns 2021.

ee. Heard 2022c, chap. 23, provides a basic outline of how the review process works, although it's oriented for the writer, not the reviewer. Among resources intended to demystify the process of writing a peer review, we like British Ecological Society 2013 and Scrimgeour and Pruss 2016.
ff. On writing a response to reviews, see Heard 2022c, chap. 24.
gg. Mlynarek et al. 2017, for example, was exactly this kind of project in Steve's lab. Heard 2018 explores the vexing question of whether, and how, a coauthored paper might appear in the graduate thesis of one or more students.

diverse perspectives on everything from healthy writing habits to where to place commas; at the same time, this varied advice can be seen as the very conflict you're trying to avoid by coordinating writing instruction across a program. Therefore, even a perfectly harmonized approach to writing instruction within a program doesn't remove the need to equip students (and mentors) to tackle conflicting advice. And realistically, getting to program-level, coordinated writing instruction is hard (understatement alert—each of us has been pushing that boulder up the hill for years in our home department), so few students will experience conflict-free advice even inside their programs. We'll wrap up with some thoughts about how to prepare students for the conflicting advice that's sure to come.

THE BEST APPROACH IS PROACTIVE

Think through likely ways that conflict will arise, in discussion with the developing writers you mentor, in advance. Have conversations about the benefit of multiple perspectives on a piece of writing. Then, together, make a plan for how to handle conflict when it happens. Here are a few ideas:

- **As a writing mentor, work to identify your own perceptions of "correct" writing.** Be honest with yourself: what do you want to change because you'd write it differently, versus what do you want to change because it *should* be different to ensure clarity, accuracy, or persuasiveness? To start, you can review the resources we recommend throughout this book and reflect on which of your writing peeves are matters of objectively correct writing, which are conventions in particular writing genres, and which are merely personal preferences (see chapter 5, "Encourage careful thought about the genre and its constraints"). Then you can make these different kinds of objections explicit to developing writers and decide how to mark them clearly when you engage with a developing writer's work. (Or, in the case of preferences, decide not to mark them at all).

- **Help developing writers to realize that they can (and are encouraged to) defend a writing choice, especially with an evidence-based defense.** That means explaining not just that they took advice from elsewhere, but explaining why they believe it was good advice. There's a consequence for mentors here: when considering a developing writer's drafts, you'll need to ask for explanations rather than "correcting" the writing. Doing so encourages metacognition and invites the developing writer into a dialogue (see chapters 5 and 6). Of course, it may also introduce more iteration and extend the timeframe over which you review a writing project. We remind you that when it's well designed, iterative feedback is powerful and needn't be as time-consuming as you might think (see chapter 3).
- **But prepare developing writers to know when to concede the point.** Writing instructors' foibles are often a matter of ego, not evidence. And that, of course, is the most tenuous but vigorously defended turf in academia (or anywhere, really). There will be occasions when you support how your student wrote something but advisory committee members or external collaborators are digging in their heels for their own writing style. *You* might be able to disregard a collaborator's preference, but a student doing the same might risk access to shared data, funding, reference letters, etc. Help students understand that even when they'd like to, sometimes they can't reject 100 percent of someone's comments without burning bridges. Remember that there's usually a third way, too: they can identify an alternative revision that satisfies someone else's edits without erasing their own writing style.

We'll close in a way we think is hopeful and forward-looking. Designing writing instruction in a programmatic framework renders explicit the foibles and complexity of writing feedback, because everyone involved with the program will have their own take on "good" writing. By working with your colleagues to resolve disagreements about writing style in advance, for the benefit of students, you'll be addressing the systemic issue. This is better than (as so often happens) making each individual student or mentor responsible for grappling with the system. But don't be misled by the impersonal word "systemic": as faculty members and other mentors, we *are* the system, or at least a big chunk of it. It's our responsibility to improve the system in ways that genuinely work for students. It's a nice bonus that the systemic reform we're talking about—like program-wide writing instruction—will also relieve us as individual mentors of the recurrent tensions that arise when writing feedback *isn't* programmatic.

Afterword

HOW DO YOU KNOW YOU'RE DOING A GOOD JOB? AND HOW DO YOU CONVINCE OTHERS THAT YOU ARE?

It might seem like we're posing two different questions here, but the answers to both rest on the same actions. In this short afterword, we'll sketch out an approach that can take you a long way toward gauging how you're doing as a writing mentor.

Before we do that, though: the fact that you're reading this book means you've taken a deeper dive into the evidence that informs effective writing instruction than most mentors ever will. Pat yourself on the back for caring enough to do that! Still, you might want more—you might want to know if you're using the ideas in this book effectively, and you might want to know how to make that evident to your students, and to the people who hire, review, and promote you.

Here's an important affirmation: evidence-based teaching or mentoring is an act of scholarship. "Scholarship" is a broader concept than "research," and it gets used more in the arts and humanities than it does in the sciences, but it's an important part of the scientific enterprise. "Scholar" means both student and teacher, both learner and learned person.[1] Some definitions of scholarship, at least, assert that it entails a combination of five dimensions: discovery, application, integration, teaching and learning, and engagement (of people beyond academia).[2] We hope you've seen all five of those in what you've read so far. Your work to mentor writing well is a kind of applied learning that should be valued—when you can demonstrate success. Indeed, recognizing yourself as a scholar connects nicely to the notion that mentoring writing, just like writing itself, isn't an innate talent but a set of skills in which you can train and grow.

As nice as that all sounds, we do need to acknowledge what this stance is up against. Making your scholarly work legible to others (students, peers, administrators, funders) is always a challenge. This can be especially true of teaching and mentoring, unless you work in disciplines that teaching is formally a part of (i.e., education, writing studies, extension services) or in institutions where teaching is the principal work (e.g., primarily undergraduate universities). Otherwise, teaching and mentoring tend to be terribly undervalued compared to publishing in peer-reviewed journals, securing external grants, or work-beyond-academia activities such as developing patentable technologies or devising policies.[3] Neither you nor we can change that value system overnight. Thus, you'll have to take your own satisfaction in an evidence-based approach to mentoring writing. You should, therefore, take the time to articulate what it is about mentoring developing writers that's meaningful to you, and to find ways to recognize when you've achieved it.

Fortunately, the path to justifying effort and demonstrating success for yourself is pretty similar to how you'll demonstrate that to others. It's also pretty similar to the process you'd use in justifying a research project! Consider that a research project begins with observations and curiosity, which lead to research questions, hypotheses, and then appropriate methods for pursuing the questions. Then, with results in hand, you use analyses to understand whether, and how, your methods succeeded in answering the questions. You can use a very similar sequence to articulate your intentions as a writing mentor and then assess the results:

- Problem statement: What problems relating to students' development as writers are you most interested in?
- Gap you can address: What dimensions of these problems are you equipped to address? How does your situation (considering the roles you play in your institution and your relationships with students and colleagues) make you so equipped?
- Proposed work and justification: How will you structure your own learning, and then your course, mentoring model, etc., to address the problems you've identified?
- Data: How will you gather data to measure your growth as a mentor and your students' growth as a result of your interventions?
- Assessment of results: How will you use the data to measure your degree of success in your own and your students' growth?

Note that this sequence is exactly how Bethann coaches developing writers to structure grant proposals. We're suggesting that you think as robustly and persuasively about what you're trying to do as a writing mentor as you would when pitching a research project for funding. Granted, you're not likely to write a hundred-plus-page funding application to teach writing. But when you can articulate the value of evidence-based mentoring, you'll find yourself better equipped to make visible the work you do and to advocate for that work to "count" in your professional setting. Using a proposal framing that your peers are familiar with also helps you do what effective writing does: calibrate to your audience.

Your plans for improving and assessing writing mentorship should be concrete, actionable, and realistic (you can think about the tried-and-true framing of SMART or EASY goals[4]). You can also draw from the many resources that exist for articulating learning objectives[5]—while these usually focus on students and courses, you can borrow from them to articulate your own goals, too. These perspectives suggest narrowing in on very specific objectives and methods for each situation in which you deal with developing writers. Imagine, for example, that you teach a content course wherein you require considerable writing (situation 1), while also directing a lab involving several graduate students and a postdoc (situation 2). We recommend specifying goals for each situation separately, at two scales: semester and academic year. These tiered goals let you, and the students you mentor, see more immediate goals within the larger framework of the growth you're targeting. Your best bet is to keep yourself focused on one to three for each period of time and each situation; however, keeping longer-term goals consistent across multiple situations can give you coherence in your efforts and the efficiency that comes with that.

Once you've set goals, you'll need to articulate what success looks like—both your own success in growth as a mentor and your students' success in growth as writers. Seeing and documenting the latter is naturally part of seeing and documenting the former. Nevertheless, we're going to focus here on your goals for *yourself*, because assessing student success in meeting goals in writing is simply a special case of assessing student success more broadly.[6]

What might this all look like? We offer an example in table A.1, for the two mentoring situations we've already mentioned. You'll notice that we've deliberately chosen a simple, accessible set of goals for someone early in their journey as a writing mentor. Does our example strike you as too easy to bother with? Great—that's evidence you're well on your way! Pick something more challenging and develop it along the same lines.

Perhaps you think this sounds like a lot of work. Perhaps it is—but you'll likely have to report to someone, at some point; having a plan and documentation now can ease that effort later. We also encourage you to keep in

TABLE A.1

Example of goals for writing mentorship and how to assess efficacy

FOR A CONTENT COURSE	FOR A LAB GROUP
Academic-year goals	
▸ Enhance my awareness of genre conventions in my discipline ▸ Improve my ability to discuss those conventions with my students.	
Semester goals	
▸ Understand that there are distinct conventions in each genre in my discipline. ▸ Effectively use conventions of one genre to inform writing and revision in that genre (e.g., Introduction section of a journal article), including structure and typical rhetorical moves.[a]	▸ Understand that there are distinct conventions in each genre in my discipline. ▸ Effectively differentiate between conventions in various outlets in a single genre (e.g., different types of journals) and between various genres (e.g., journal articles, technical reports, scicomm pieces). ▸ Learn about writing conventions in a genre I haven't previously written. (Interpret this broadly, choosing a genre that interests the lab group but that's far from anything you've done before—consider a policy brief, a podcast script, or museum visitor guide.)
What would constitute success?	
▸ I can explain at least three commonalities and differences between typical genres in my discipline, taking this beyond surface level to include structure, sequence, and rhetorical moves. ▸ I can provide feedback to a writer regarding their use of these conventions and articulate how their later drafts demonstrate improvement.	▸ I can analyze and articulate differences between writing conventions in three different journals (e.g., *Science*, a discipline-focused specialty journal, and a discipline-focused applied journal). ▸ I can draft (perhaps jointly with the lab group) a detailed table analyzing commonalities and differences between typical genres in our discipline—including the new genre I tackled. ▸ I can identify and discuss effective and ineffective use of conventions in a developing writer's draft text.
Methods of documentation[b]	
▸ Self-study (At the start of the semester, write a description of general genre conventions and of conventions for a target journal. Repeat at the end of the semester, then compare with the initial description.) ▸ Improvement in assignment prompts and lesson plans[c] ▸ Colleagues' observations[d] ▸ Student responses to midterm and end-of-term feedback surveys[e]	▸ Self-study (At the start of the semester, write a description of general genre conventions, conventions for a target journal, and for writing in the new genre you've chosen. Repeat at the end of the semester, then compare with initial description.) ▸ Reflections on efficacy, learning, and things to change next time ▸ Calendar of topics for lab discussions, with reading lists (Track how this list matures with time.) ▸ Writing development plan for the lab group

Actions to achieve these goals	
In addition to action items in the methods column, I will: ▸ read Heard's chapters on genre conventions and select three to five conventions to discuss with students.[f] ▸ introduce genre conventions to students using examples from popular media (press release, social media, newspaper, etc.) ▸ assign students to compare conventions in articles from two or three types of journals ▸ assign students to compare two or three genres relevant to class, using vocabulary regarding conventions ▸ assign students to write reflections self-assessing efficacy and growth in this area	In addition to action items in the methods column, I will: ▸ read and discuss Heard's chapters on genre conventions with the lab group ▸ discuss genre conventions using examples from popular media (press release, social media, newspaper, etc.) with the lab group ▸ discuss with at least three colleagues how they foster writing growth ▸ task students to individually compare conventions in articles from two or three target journals relevant to their work ▸ task students to trade writing and provide each other feedback about effective use of genre conventions ▸ task students with self-assessing efficacy and growth in this area (lab discussion or individual, written reflections)

a. While most of us have an intuitive idea of what goes into an Introduction, many of us struggle to enunciate that in the concrete way that helps students produce an Introduction meeting reader expectations. It will be helpful to go beyond notions of the "narrowing part of the hourglass" to consider the pattern of rhetorical moves that's typical in an Introduction. For an overview, see Heard 2022c, chap. 10; for deeper dives, see Swales 1990, among others.
b. For rationale on collecting a wide range of evidence to inform your assessment of writing mentorship, see Jackson 2015. We caution that although it's an obvious possibility, it's probably unwise to rely on student grades as evidence of success. Dayton 2015 and Jackson 2015 discuss the case for decoupling grades from assessment of instructor efficacy.
c. Compare how you describe assignments or plan lessons now versus before you read this book and/or did other work to improve your ability to mentor developing writers. For instance, you could look at how an assignment is described, and document how its framing has become more concrete and specific, how you're using evidence-based vocabulary now, etc.
d. You may appreciate and trust feedback from someone you know well. However, evidence indicates that observations of teaching are reliable only if the observer has been trained in advance *and* if there are explicit and clear evaluation criteria (see several chapters in Dayton 2015). Furthermore, observers have ideally calibrated their rating process in advance, and instructors should be evaluated by at least two people with multiple observations throughout the term. See Dayton 2015 for a detailed discussion, and Jackson 2015 for an example of how to design observation criteria. The obvious caveat is that multiple observations by multiple observers per term are often infeasible, making one-off observations inevitable; thus, caution should be applied to their results.
e. Bethann has discussed in detail (Merkle 2024a) how she uses students' start-of-term goal statements, midterm feedback, and end-of-term goal reflections to inform her understanding of course efficacy, and how these can be used in annual reviews and the like. Anson 2015 offers another useful perspective on how feedback can be collected, including the helpful reminder that such feedback should concentrate on what you want students to be capable of, not content you covered. (We are ambivalent, however, about his recommendation to make public the results and associated instructor commentary.) Note that this is not an invitation to place weight on standardized student opinion surveys; student course evaluations are well known to correlate poorly with learning and to be biased against instructors belonging to various marginalized demographics (see, e.g., Jackson 2015, Moore 2015, Mengel et al. 2019, Hessler et al. 2018, Dakota Murray et al. 2020, and Sophie Adams et al. 2022). For an interactive dataset with specific, verbatim examples, see Ben Schmidt's data visualization, Gendered Language in Teacher Reviews (http://benschmidt.org/profGender/).
f. Heard 2022c; or of course use an alternative guidebook (see the appendix).

mind that you wouldn't balk at producing a comprehensive management plan, budget, objectives, and planned methods for a research project. If you want to drive the bus regarding the assessment of your writing mentorship, you'll need to approach that with rigor, too. But don't panic; doing *some* self-assessment is better than doing none. If you're about to start the current semester, or you're partway through it, you can choose one or two elements from the table to focus on, or take a simpler "just in time" approach for now and plan for a more thorough assessment next time around.[7] We encourage you to do at least that.

Once you've done all this work and collected all this evidence, what do you do with it? You likely have at least three audiences that should see and consider what you've done:

- **Yourself.** Work from affirmation, not for it,[8] remembering to take satisfaction in your accomplishments. It often takes explicit self-assessment to see those accomplishments instead of wallowing in the self-doubt or self-recrimination that you *still* haven't done "enough."
- **Your students.** They are the people whose writing growth you're working to enhance. They'll appreciate knowing about the work you're doing to improve your own mentoring and thus their development as writers. Not only that, you'll be setting them up to be conscious of their own development when they become mentors. They may also be surprised to know that they're already mentors themselves (perhaps with their classmates, perhaps with younger friends or family members). They may not yet have thought much about how important such roles will be for them in the future, and your reporting back to them can help.
- **The people who decide your professional future.** Such people might include a hiring or tenure-review committee, an administrator, or someone else. You're in a strong position when you can convey clearly what you're working to improve, how you're working to improve it, and how you know you've made progress. Include language in applications, review materials, etc., that articulates why you value evidence-based writing instruction and mentorship. When you do, cite the peer-reviewed literature informing your stance (most people reviewing you are accustomed to knowledge claims being referenced). Depending on the format, you may also include materials such as syllabi, lesson plans, anonymized student comments. A combination of these types of evidence is likely to be most compelling.

We'll close with a reminder: as we've noted a few times, there's no threshold beyond which someone is done developing as a writer. The same is true about developing as a mentor. Your interest in this book suggests you're on the journey. Know that you're not alone in caring about effective mentorship of developing writers. Help your mentees and your colleagues join you, and keep an eye out for your fellow travelers. We'll see you on the road!

Acknowledgments

We thank our editor, Mary Laur, for believing in our book from the beginning. We thank the production staff who have done such great work with the manuscript, especially copyeditor Johanna Rosenbohm, Nick Lilly, Lindsy Rice, Beth Ina, Laura Leichum, indexer Alexandra Peace, and Andrea Blatz. We thank several anonymous reviewers for helpful comments, and we especially appreciate our beta readers: Priya Shukla, Jessie Rack, Laura Portwood-Stacer, Terry McGlynn, Lesley Fleming, and Pauline Barmby. Johanna Varner, Albrecht Schulte-Hostedde, and Kristin Benson made helpful comments on elements of the text; and several folks, especially Jennifer Purrenhage and Riley Bernard, were willing to test-drive ideas and resources that became part of this book. We thank Ben Dow for helping wrangle the references. We thank readers of our blogs and our social media communities for answering questions and making suggestions about mentoring writing. Among other things, they pointed us to vital, intersectional ideas, scholars, and movements that have shaped the perspectives on inclusive pedagogy that underpin this whole book.

We are each grateful to the students, authors, editing clients, and other developing writers we've worked with. Yes, occasionally we were frustrated (as were they, we're sure!); but we learned from them, and our interactions reinforced, every time, that we can all get better at writing, and that compassionate help and feedback can make all the difference.

Kelly Kinney hooked Bethann on writing studies and cheered her on as she envisioned and pitched this book (including to Steve!). Nancy Small, Peyton Lunzer, Francesca King, Anne Grass, Tim Etzkorn, and Alex Cargol supported Bethann's work on this project in myriad, quiet ways. Special thanks to Rick Fisher, who Bethann calls "the best co-instructor I've ever worked with" and who is an ongoing source of knowledge and inspiration and a partner in providing resources and support programs for developing writers and their mentors. This book wouldn't exist without the work Rick and Bethann have done together. Jeremy Fox and Meghan Duffy (of the *Dynamic Ecology* blog) hosted a set of blog posts that helped Bethann build the momentum to actually write this book.

We are grateful to the Wyoming Institute for Humanities Research for supporting indexing costs, and to the University of Wyoming College of Agriculture, Life Sciences, and Natural Resources' Global Perspectives fund for supporting Bethann's travel to New Brunswick to work with Steve on the manuscript at a critical stage.

Finally, we thank our families, who have always affirmed the rabbit holes we run down while trying to make the world a better place. Especially, Bethann thanks her parents, who taught her to love words, her sisters for their support, and her husband, Jerod, who has listened to more rants and iterations of the advice in this book than anyone, yet remains genuinely supportive. Steve thanks his wife, Kristie, and their son, Jamie, who were so patient as he worked on this book.

Appendix: Resources for Writing Teachers/Mentors

We love that you're reading this book, and we think it can help you a lot in your quest to mentor scientific writers more efficiently and effectively. But there are other resources you can lean on, too. These will help you find different perspectives, dig deeper into the issues that are most important to you, or find writing exercises you can use in classes and beyond.

Here we provide a list of books and other resources that we recommend. We know you won't want to tackle all of them, so we annotate our list, telling you the strengths and ideal audiences of each. We've also marked good sources of writing exercises with an Ⓔ.

An online version of this appendix can be found at https://www.helpingstudentswrite.com/resources. We'll keep the online version updated; if your favorite resource doesn't appear on our list, you can recommend that we add it.

Part 1: Resources for teachers/mentors

BOOKS

These resources are aimed primarily at teachers or mentors, mostly but not exclusively of writing. Few of these are specific to science, and most emphasize the undergraduate classroom simply because that gives them the largest market. You can almost always extrapolate to other mentorship situations.

Our top recommendations

The Chicago Guide to College Science Teaching by Terry McGlynn (University of Chicago Press, 2020). McGlynn tackles undergraduate teaching in general, but much of what he has to say is relevant to teaching writing. He covers construction of a syllabus, methods for classroom teaching and behavior for teachers, assigning and assessing assignments and exams, online teaching, and more. Empathy and respect for students is a major emphasis. While McGlynn's advice is underpinned by peer-reviewed literature in the scholarship of teaching and learning, it's delivered plainly, as one scientist to another, and builds on his direct experience in science classrooms.

Ⓔ *Engaging Ideas: The Professor's Guide to Integrating Writing, Critical Thinking, and Active Learning in the Classroom* by John C. Bean and Dan Melzer (Jossey-Bass, 2021). This is *the* evidence-based guide to mentoring writing in the undergraduate classroom. While it works from a deep grounding in writing studies and in rhetoric and composition, it's approachable and addresses a wide range of disciplines (including the sciences). It covers everything from designing a course, setting expectations, assessment, and giving feedback on writing assignments to student-forward approaches such as small-group workshops and peer review. Plus it can be sampled to address specific issues without requiring a full read-through.

Radical Hope: A Teaching Manifesto by Kevin M. Gannon (West Virginia University Press, 2020). This book is a brief, valuable, and forward-looking discussion of what makes teaching so hard, and how you can pick your battles in ways that makes teaching more effective and meaningful.

More great choices

Ⓔ *Assessing the Teaching of Writing: Twenty-First Century Trends and Technologies*, edited by Amy Dayton (Utah State University Press, 2015). While we didn't have room to spend much time on this in the book (except in the brief afterword), you might be reasonably concerned that you're going to get pushback as you adjust how you teach scientific writing. You're working for change, which a lot of people instinctively resist (students and colleagues alike). It will help to have a concrete plan for how you're going to document and assess the efficacy of your shifts in teaching. This book provides ideas for how to do that in ways that can be legible to students, colleagues, and administrators.

Happier Hour: How to Spend Your Time for a Better, More Meaningful Life by Cassie Holmes (Penguin Life, 2024). While this one might seem tangential, it's quite possible that you're mentoring writing in a setting where virtually any other activity is more highly valued. Even at a primarily undergraduate institution, you may find that the R1 academic paradigm seeps in (this is calibrated to prioritize research productivity; the term comes from usage in the United States). If caring about things like teaching makes you an outlier, you'll need to articulate your own priorities and the satisfaction you derive from being a good teacher and mentor. This book and Kevin Gannon's *Radical Hope* (above) make a good set to help you with that, even in a publish-or-perish environment.

Minds Made for Stories: How We Really Read and Write Informational and Persuasive Texts by Thomas Newkirk (Heinemann, 2014). This book is a valuable resource for anyone supporting developing writers who would like to enhance the persuasive dimensions of their writing. Newkirk tackles many of the standard assumptions about academic writing, then provides mentors with specific reframes that can be used in the classroom. (While grant proposals, job applications, and the like are explicitly persuasive, *all* successful academic writing relies on persuasion. Understanding this can afford significant growth to writers.)

Ⓔ *Teaching Science Students to Communicate: A Practical Guide*, edited by Susan Rowland and Louise Kuchel (Springer, 2023). This edited collection is jam-packed with evidence-based lesson plans you can use in planning a scicomm assignment or course. It has numerous activities to provide practice in writing and in complementary skills. For example, Bethann has a chapter in this book that we've cited several times—it deals with helping students plan, manage, and complete a major, semester-long communication project. The process she details can be readily used to help new students think through a complex, multistage writing project like a thesis or peer-reviewed manuscript.

They Say / I Say by Gerald Graff and Cathy Birkenstein (W. W. Norton, 2021). This book breaks down a lot of the typical academic-writing "moves"—rhetorical strategies such as hedging, disagreeing tactfully, articulating a need for a specific research question, etc. Technically, the book is written for developing writers, but we listed it here because it's also an enormously helpful guidebook for mentors: it provides you with word-for-word templates and the vocabulary you need to take your instincts around writing and turn them into concrete examples and rationales to share with developing writers.

Why Students Resist Learning: A Practical Model for Understanding and Help-

ing Students, edited by Anton O. Tolman and Janine Kremling (Routledge, 2016). This book examines reasons why students resist active learning—which is what you'll be asking them to do if you follow the advice in this book. If you're aware of the evolutionary, social, and systemic reasons why students resist or struggle to do what we ask, you can help them without taking their entirely predictable resistance personally.

Writing in the Sciences: Exploring Conventions of Scientific Discourse by Ann M. Penrose and Steven B. Katz (Pearson, 2009). This book might seem like it might be better as a text for developing writers. But we've listed it here because it does some of what our book does: demystifying scientific writing by leveraging the disciplines of rhetoric, composition, linguistics, psychology, sociology, philosophy of science, and science education. (The authors are professors of English who study the rhetoric of science.) If you're looking for vocabulary and concepts that make explicit what we do as scientific writers, this book is a great starting point.

Writing Spaces: Readings on Writing (Parlor Press, 2010–24; https://writingspaces.org/). This book series is an open-access compendium of peer-reviewed essays on writing, generally intended to be read by undergraduate students. So why are we listing it under "Resources for teachers/mentors"? Because it's huge (over 100 essays so far); because it covers a lot of ground (from punctuation to rhetoric to peer review to PowerPoint design); and because relatively little of that ground is directly focused on STEM writers or writing. You'll find it a valuable source of essays you can give to your students, but we think sending students to it directly would leave them overwhelmed.

ONLINE RESOURCES

Both of us write blogs that frequently address writing. We are typically talking to you, but you might also find some gems in our blog that you'll share with students.

Scientist Sees Squirrel (https://scientistseessquirrel.wordpress.com/). This is Steve's blog, and while it's not particularly focused (hence its title!), it frequently tackles issues of writing, and of mentoring writing.

School of Good Trouble (https://www.commnatural.com/blog). This is Bethann's blog—and rather like Steve's, it's often about writing or mentoring writing; but it ranges pretty far from that, too.

Beyond our blogs, here are some additional valuable resources.

Cult of Pedagogy (https://www.cultofpedagogy.com/). While this blog and collection of resources is aimed mainly at K–12 teaching, there are lots of ideas here that can be useful in higher education, too. It's about teaching in general, so you may want to search on "writing."

Faculty Focus (https://www.facultyfocus.com/). This newsletter on teaching in higher education ranges more broadly than writing, but a quick search on "writing" will return lots of advice and resources, from multiple contributors.

Megan McIntyre (https://www.meganmmcintyre.com/resources). McIntyre directs the Sonoma State University Writing Program and has compiled resources and advice sheets for teaching university-level writing. Her resources are particularly strong for course design.

Metawriting (https://metawriting.deannamascle.com/). Deanna Mascle is director of the Morehead Writing Project at Morehead State University. She describes herself as a writing evangelist, but it would be just as accurate to think of her as a teaching-writing evangelist. Posts tend to be short and focused on a particular idea for a writing exercise or writing pedagogy practice.

Part 2: Resources for writers

These resources are primarily aimed at writers rather than those who mentor them. However, we include them here because as a mentor, you can use them in at least three ways. Most obviously, they can be assigned resources ("textbooks") for a writing class or a lab (or other) group. Second, you can recommend them on an individual basis to writers you mentor, based on their stage of development and particular needs. Third, they can help you understand conventions and best practices in scientific writing (and why those conventions have come to exist), so you can provide evidence-based advice and assessment to your students.

BOOKS

We'll start with books. There are many, many books for writers; even the small subset we list here tried hard to get completely out of control. As a guide to tackling our list: the first two are our top two. If you want to take advantage of a broader set of resources, look beyond those for a few that meet your (or your students') particular needs; or consider making a bookshelf of them available to your students.

Our top recommendations

Ⓔ *The Scientist's Guide to Writing: How to Write More Easily and Effectively throughout Your Scientific Career* by Stephen Heard (Princeton University Press, 2022). This book offers in-depth treatment emphasizing the journal paper, so those intending academic careers will find it most on target (however, nonacademic science writers needn't do much extrapolation). It covers expectations for the *form* of scientific writing (explaining the historical and functional reasons for the conventions we follow, using examples across STEM disciplines), and also provides a wealth of advice for writing *behavior*. The book includes numerous writing exercises, suitable for use in assignments or workshops or for a reader to work through on their own. As a bonus, its footnotes are hilarious.

Ⓔ *Writing Science in Plain English* by Anne Greene (University of Chicago Press, 2025). Greene tackles one of the biggest temptations facing scientific writers: to write jargon-laden prose of impenetrable complexity. The book is succinct, affordable, and accurate, and provides useful examples across a range of science disciplines. Additionally, Greene provides numerous writing exercises. However, we find this book most useful when paired with another book that builds further on the writing behavior and genre-specific goals you're working toward.

More great choices

Ⓔ *Academic Writing for Graduate Students: Essential Tasks and Skills* by John M. Swales and Christine Feak (University of Michigan Press, 2012). Because Swales and Feak write for students in *any* academic field, science graduate students may find this book harder to relate to. However, it's built on deep understanding of how academic texts are constructed and is strong on the rhetorical patterns academics use. This book provides abundant writing exercises (defined very broadly, including such allied tasks as specifying hypotheses, searching literature, and reading papers).

Bird by Bird: Some Instructions on Writing and Life by Anne Lamott (Pantheon, 1994). This is a pocket-size book that delves into the mindset and habits of successful writing. When the developing writers you support are ready to take their writing to a level of sophistication and self-expression beyond technical correctness, this book can support (and entertain) them.

English for Writing Research Papers by Adrian Wallwork (Springer, 2011). This is a book for those writing in English as an additional language (see our chapter 8). It includes both general advice about structuring scientific papers and more focused advice about English grammar and syntax, especially in areas that often trip up EAL scientific writers. An alternative is *Science Research Writing: For Native and Non-native Speakers of English* by Hilary Glasman-Deal (World Scientific Publishing, 2020).

Getting to the Heart of Science Communication: A Guide to Effective Engagement by Faith Kearns (Island Press, 2021). One of the few books available about *scicomm* rather than strict-sense scientific writing, Kearns's book is the go-to reference for evidence-based approaches to science communication. Developing writers and mentors alike can appreciate this book for its advice, its connections to literature informing scicomm, and its treatment of the ethics of scicomm. You can think of this book as the scicomm equivalent of scientific writing–advice books like Steve's.

Handbook of Writing for the Mathematical Sciences by Nicholas Higham (Society for Industrial and Applied Mathematics, 2020). Of all the STEM disciplines, mathematics has the most divergent conventions for writing (that's especially true for pure mathematics). Higham's guide covers writing and publishing papers and also offers advice preparing posters, talks, and books. He includes tips for use of LaTeX, the typesetting software used by most mathematical writers.

How to Write a Lot: A Practical Guide to Productive Academic Writing by Paul Silvia (APA LifeTools, 2018). This is a short book about writing (the verb), not writing (the noun). Silvia focuses on writer behavior, especially around time management and the psychology of productivity. We know few scientific writers, developing or otherwise, who wouldn't like to "write a lot" and do it more easily.

The Little, Brown Handbook by H. Ramsey Fowler, Jane Aaron, and Michael Greer (Pearson, 2022). This is a reference manual to grammar, composition, and rhetoric. It's dull, but it's valuable as a way to answer questions about punctuation, word usage, verb tenses and moods, and similar things about the way the English language works. There are several similar guides; this one is comprehensive and straightforward to use. Cheaper older editions are just as good.

On Writing: A Memoir of the Craft by Stephen King (Scribner, 2020). Yes, this is a book by a horror novelist, but it has a lot to say to writers in any genre—including scientific writers. King gives a lot of good advice about why to care about writing as a craft, and how to take writing seriously (as one needs to). It's simple and very engaging.

The Science Writer's Essay Handbook by Michelle Nijhuis (pub. by author, 2016). *The* resource for popular science writing, Nijhuis's book is short (just 83 pages). This makes it a great reference for developing writers looking to write beyond scientific audiences. (This book doesn't get into the business side of being a science writer—see our suggested websites below for that.)

Ⓔ *Scientific Writing and Communication: Papers, Proposals, and Presentations* by Angie Hofmann (Oxford University Press, 2022). Hofmann's book takes a more comprehensive, almost encyclopedic approach to scientific communication, covering everything from research applications to job applications. Because the resulting book is long and somewhat dense, it's unlikely that you can expect a mentee to read it through; still, it's worth having on a lab bookshelf as a reference. Hofmann provides both end-of-chapter exercises and self-assessments that can also be useful in mentoring.

Ⓔ *Storyworthy: Engage, Teach, Persuade, and Change Your Life through the Power of Storytelling* by Matthew Dicks (New World Library, 2018). While this sounds a bit like a self-help book, it's more than that: it can help a developing writer understand the power and utility of anecdotes, narrative tools such as story arcs (even in technical writing), and writing in an approachable-yet-authoritative tone. This book can help students produce more engaging cover letters; practice for conference posters, presentations, and interviews; or even frame effective grant proposals, press releases, and more. It's loaded with exercises that developing writers can tackle on their own, or that you can adapt for classes.

Stylish Academic Writing by Helen Sword (Harvard University Press, 2012). Sword, like Greene, suggests that academic writing needn't be dull. This short book explores, with examples, techniques for making writing clear and engaging—even stylish and vivid. This book will be most valuable for developing writers who have the basics in hand.

Writing Science: How to Write Papers That Get Cited and Proposals That Get Funded by Joshua Schimel (Oxford University Press, 2011). This book is older than some of our other recommendations, but its main focus on building and communicating a coherent story is timeless. We think *Writing Science* will appeal more to graduate students than to undergraduates.

ONLINE RESOURCES

There are also web resources for writers—*many* of them. Here are a few you or your students may find useful.

Academic Writing Amplified (podcast; search for it wherever you get your podcasts). Cathy Mazak's podcast is aimed at helping women and nonbinary academics with writing challenges, but most of the content will help any academic. Mazak's background is in the scholarship of teaching English, but she is broad-ranging as a writing coach. Early episodes are particularly strong around issues of writer behavior.

Morley's Academic Phrasebank (https://www.phrasebank.manchester.ac.uk/). This is a compilation of thousands of typical phrases from academic literature—for example, alternative ways of expressing the level of confidence you might have in a result. A caution, though: the phrasebook lists *typical* academic phrasing, which means many of the phrases are dreary, passive, or hackneyed—just like our literature. Users of this resource should use it to find phrasing that's common, but then think about whether they can improve on it.

Raul Pacheco-Vega's blog (http://www.raulpacheco.org/blog/). Pacheco-Vega is a social scientist (geographer and political scientist) whose blog ranges widely. He frequently writes about research methods (mostly for qualitative research) and academic writing, and his advice about productive workflows and habits is valuable.

The Open Notebook (https://www.theopennotebook.com/) and the National Association of Science Writers (https://www.nasw.org/). These are the first stops on the web for anyone looking to do popular science writing as either a hobby or part of their profession. Both provide extensive advice (and a professional community) around the business and craft sides of science writing.

Shut Up and Write (https://shutupwrite.com/). For writers who need motivation or accountability to get down to the business of writing, *Shut Up and Write* provides access to online writing groups. Mostly, these are organized around sessions in which people write together (not coauthoring, just writing at the same time) rather than discuss their writing. Writing discussions could, however, be a spinoff from a *Shut Up and Write* session.

The Thesis Whisperer (https://thesiswhisperer.com/). This is Inger Mewburn's blog, and it's full of helpful advice on both writing behavior and academic-writing style.

The Writer's Diet (https://writersdiet.com/). This is a set of tools, from Helen Sword, for making writing simpler, clearer, and less stodgy. Although many will find the diet metaphor problematic, the tool is valuable because it targets one of the most pernicious of our scientific-writing habits. (There are alternative metaphors available via the web tool.) There's an accompanying book (Sword 2016b), but the website is free. The BASE diagnostic tool (https://writersdiet.com/base/about/) is our favorite. It asks writers to reflect on their habits in four key areas characteristic of successful, productive writers: behavioral habits, artisanal habits (which Bethann sometimes frames to students as attitudinal habits), social habits, and emotional habits. Students get a *lot* out of being prompted to consider that there is more to writing than struggling alone.

Notes

INTRODUCTION

1. See Fox and Merkle 2019 for some (informal) poll data suggesting that if you think you should teach writing more, but don't see how you possibly can, you're in good—or at least extensive—company.
2. By a "process" we mean a workflow or an approach to a task; by a "skill" we mean the ability to execute a process efficiently and well. Writers develop many processes—one for scheduling writing work, one for outlining sections, one for proofing citation lists, and many, many more. Each involves skills. We'll have more to say about most of these! However, it's a bit overwhelming to *start* a book waist-deep in detail, so we're just pointing to these ideas for now.
3. It isn't the aim of this book to help *writers*, or at least, not directly. But don't despair: if you find scientific writing hard, there's help available. Steve, unsurprisingly, suggests you start with his own book: *The Scientist's Guide to Writing* (Heard 2022c). There are many other good resources out there, and we provide some annotated recommendations in the appendix. Grogan 2020 also helpfully distills what's hard about scientific writing and offers a few key tips to compensate.
4. What's "scaffolding"? Scaffolding in teaching operates like scaffolding in construction: providing temporary structure to assist step-by-step building work. In teaching, scaffolding is a sequence of stepwise and often iterative tasks that break down a learning process or project into increments. Those tasks may include drafts with feedback and revision but aren't limited to that; scaffolding can also build in interactions with theory, literature, examples, and more. We might scaffold something like a major writing project because students benefit from incremental help as they work to recognize the complexity of executing a large project. Over the course of a semester or other learning experience, scaffolding should also support students in gaining the skills needed to deal with that complexity on their own. In chapter 4, we provide an example by detailing a sequence

of assignments building up to completing a major project. For useful reviews of scaffolding in pedagogy, see Belland 2014, Boblett 2012, and Benson 1997. In chapter 10, we discuss the idea of scaffolding across entire curricula.

5. We use the word "assessment" here where you might have expected "grading." There is considerable discussion around best practices for what mentors do with submitted work. Many still use traditional grading. Others use variants such as contract grading or standards-based grading; still others give only qualitative feedback, not grades, until the end of the course (this approach is most often what people mean by "ungrading"). We use "assessment" throughout this book to capture this variety of approaches.
6. For example, we're both active in popular as well as scientific writing. Steve's blog is *Scientist Sees Squirrel* (https://scientistseessquirrel.wordpress.com/), and his book *Charles Darwin's Barnacle and David Bowie's Spider* (Heard 2020a) is an exploration of species naming for anyone curious about nature, history, and personality. Bethann has written three-hundred-plus popular articles for outlets like *American Scientist* and *Western Confluence* and maintains the *School of Good Trouble* blog (https://www.commnatural.com/blog) as well as a podcast, *Meteor Scicomm* (https://www.meteorscicomm.org/podcast). Finally, it's worth mentioning that despite C. P. Snow's (1959) famously misunderstood gripe that science and the arts are two cultures erroneously divided, scientists also write novels (e.g., Nabokov 1955, Sagan 1985), poetry (e.g., Anand 2022, Conn 2023), and more. While teaching creative writing isn't a focus of this book, we do make some remarks about transferability of skills; see also Januchowski-Hartley et al. 2018.
7. For some general advice about respectful assessment, see McGlynn 2020. For helpful perspectives on relationships with students (with bearing on assessment), see Gannon 2020. For advice more focused on the teaching of writing, see Bean and Melzer 2021. For some discussion of "ungrading," see chapter 4, note 37.
8. We aren't dismissing the usefulness of such books! We provide an annotated list of some of our favorites in the appendix. But they don't tell you how to mentor writing; instead, they tell developing writers how to develop themselves.
9. The literature on teachers using research to inform their teaching is extensive and mixed, and—when candid—acknowledges that "the collection of research on teaching appears as a fragmented and sometimes even conflicting whole which it is difficult to adapt to practical teaching situations" (Pitkäniemi 2010, 158). In other words, if you find this challenging, you're in good company.
10. If you're looking for bracing reading on the intersections between academia, identity, writing, and self-preservation, we have drawn insight and encouragement from both the peer-reviewed and popular writing of Beronda L. Montgomery (2021, about working from, not for, affirmation in academia), Chanda Prescod-Weinstein (2021, about balancing science expertise and advocacy in graduate school), and Kevin Gannon (2020, about teaching in higher education in the face of the many reasons an educator might despair). There's also extensive literature on the intersectional issues that play into lack of respect (from students, peers, and administrators) for instructors who identify with marginal-

ized demographics. You can start with Mowatt 2019, Haynes et al. 2020, Myrtle Bell et al. 2021, Settles et al. 2021, and García Peña 2022. Misra et al. 2011 and Deanna et al. 2022 also address overwork and service burdens upon minoritized faculty and others.

CHAPTER ONE

1. These mentors (earlier in his career, Steve was one of them) might be nonplussed by the title of Helen Sword's book *Writing with Pleasure* (2023). Sword argues that writing, including academic writing, can and should be a source of pleasure. She makes extensive suggestions about how that can be achieved, although not all writers (or mentors) will be convinced.
2. Advocates and providers serving students are increasingly using the term "learning difference" (instead of "learning disability" or "disorder"). As Elizabeth Ross (2018) puts it: "Students need a term that encourages them to understand how they learn differently; instead of just assuming that they are 'worse' when it comes to school work."
3. See Boucher et al. 2013, Wilder 2013, Ideland 2018, Docot 2022, and Singh 2022.
4. Just as a start: what we used to think of as nontraditional students (over 24 years old, working full time, caring for dependents, living off campus, part-time enrollment, with a high school equivalency diploma, etc.) are increasingly the norm in undergraduate programs. In the United States, for example, nontraditional students make up around three-quarters of the undergraduate-student population (Radford et al. 2015). Around one-third of US students are currently first-generation (their parents didn't attend college; Cataldi, Bennett, and Chen 2018), and a significant proportion are from low-income backgrounds (e.g., 20–50% of students experience food insecurity, and the proportion is rising; Nikolaus et al. 2020, Nazmi et al. 2019). More students are coming from historically excluded ethnic and cultural demographics, as immigrants, with first languages other than English (de Brey et al. 2019), and with 2SLGBTQIA+ identities (O'Neill et al. 2022). There are intersectional implications of this, with higher and rising rates of poverty and risk of dropping out for female, non-White, or first-generation students, students who are first- or second-generation immigrants, or students who have English as an additional language. Even with high attrition rates as students progress through STEM training, these diverse students are becoming more prevalent in graduate programs and as postdocs and early-career scientists. This portrait of the undergraduate population stands in some contrast with the population of likely writing mentors. Of the 1.5 million faculty in the United States in 2018, 73% were White, and the majority were men (NCES 2023).
5. In addition to our discussion of these issues in this volume (see chapter 8), Heard 2022c, chap. 29, reviews such resources, while Hanauer and Englander 2013 explores the history and current status of English as the dominant language of scientific communication, and discusses issues facing EAL writers and interventions that can help.
6. Useful perspectives on prestige dialects in academic writing and STEM com-

munication include Rose 1985, G. Hull and Rose 1989, and Trachtenberg 2018. Several chapters in Bird et al. 2019a offer suggestions about how to use reflection to foster writing development for students from diverse cultural and linguistic backgrounds.

7. Here are a few starting points: T. Smith and Kirby 2021, and three collections of web resources: notes on creating a neurodiverse campus from the Neurodiversity Network (https://www.neurodiversitynetwork.net/creating-a-neurodiverse-campus), a diverse set of neurodiversity-themed resources from the Neurodiversity Hub (https://www.neurodiversityhub.org/resources-for-universities), and resources for those who teach neurodiverse students from the Heinemann imprint of Houghton Mifflin Harcourt (https://blog.heinemann.com/neurodiversity-resources-for-educators).
8. For a host of examples of how to foster your students' willingness and ability to write about themselves, as a way of building writing skills, see Bird et al. 2019a.
9. Bong and Chen 2021 reviews relevant literature and suggests practices you can use to ensure digital accessibility.
10. Shohei Ohtani and Vladimir Guerrero Jr. certainly have innate baseball abilities that neither Bethann nor Steve possess; similarly, there really are some innately "talented" writers. Consider, for example, Barbara Cartland, who published over 720 books during her lifetime, beginning at age 19, and left another 160 or so complete manuscripts upon her death. Sure, most of those books were somewhat formulaic romance novels; but there's no way Cartland could have written 720 books without something we'd reluctantly identify as genius. However, she was an outlier; vanishingly few Barbara Cartlands will show up among our students or collaborators (Heard 2015b).
11. See, for example, Alexander and Onwuegbuzie 2007, Sanders-Reio et al. 2014, and J. D. Williams and Takaku 2011.
12. Turbek et al. 2016 (which also offers a helpful overview of scientific writing for undergraduate students).
13. Sanders-Reio et al. 2014 and Pajares 2003 are good reading on this, while more broadly, research and resources on growth versus fixed mindsets may be helpful for you and the writers you support. Carol Dweck is the go-to scholar on these concepts, and Dweck 2016 offers a useful primer. Tremain 2023 addresses writerly identity from the perspective of discourse communities (e.g., the lingo of baseball fans versus molecular biologists). Bird et al. 2019a and Yancey et al. 2014 offer useful discussion of how identity plays into writing and how writing mentors can account for it.
14. And we do—both of us are prolific writers across a wide range of genres. We wouldn't write that much, in all those various ways, if writing wasn't satisfying for us. Of course, the work of getting from an idea to a polished text that does what we intend is still hard and time-consuming. In fact, it's so hard and time-consuming that we surely wouldn't draft, and revise *and revise and revise*, so much if we didn't get pleasure out of it. Helen Sword (2023) has more to say about finding pleasure in writing.
15. See Fisher et al. 2024.
16. Among writing guides, Swales and Feak 2012, Silvia 2018, Belcher 2019, and es-

pecially Heard 2022c put a significant focus on skill development. Tremain 2019 details readings for focusing readers on the growth aspects of writing. Stephen King's *On Writing: A Memoir of the Craft* (2000) might seem like an odd recommendation, but a writer of horror novels has more than you'd think to teach a developing scientific writer. (Connecting developing writers to genres that may feel more familiar and approachable is a great strategy. So is shifting students' focus away from the spooky, capital-*W* Writing and toward tasks that just happen to involve writing; Merkle 2023c). If you don't want students to tackle an entire writing book, there are blog posts and essays that might be more digestible appetizers. For example, Heard 2021 discusses the value of practice writing, Pineda 2020 makes an appeal for mindful reading, and Merkle 2023b and Mewburn 2021 outline approaches to project management. Helen Sword (e.g., 2017, 2023) offers extensive advice that emphasizes the pleasure and positive sides of the writing process. See the appendix for more recommendations.

17. This awareness gain is an important outcome of the writing courses that both Bethann and Steve teach or co-teach. Students consistently report that they learn a great deal from each other about how to manage their time, overcome fears of asking for feedback, and more. Bethann assigns both undergraduate and graduate students a reflection based on completing Helen Sword's BASE activity to identify their own writing habits. You can do the same by exploring Sword's online tool (https://writersdiet.com/base/about/) and/or assigning the relevant section of Sword 2016b. (For a detailed definition of reflection, see note 19, below.)
18. If you're looking to formalize this support, see Inger Mewburn's many blog posts about the "Thesis Bootcamps" she co-runs (https://thesiswhisperer.com/?s=thesis+bootcamp). See also Fisher et al. 2024 about how Bethann and her collaborator run the University of Wyoming scholarly writing mindset program to elicit these kinds of realizations by students.
19. On "scaffolding," see the introduction, note 4. "Reflection" and "metacognition" have become both buzzwords and pedagogical expectations (for instructors and students), in part because most scholars of pedagogy agree that they are vital (e.g., Ryan and Ryan 2015, Silver et al. 2013). However, as Rodgers 2002 discusses in detail, reflection is also a fuzzy notion unless we are very clear on what we mean by it and what role we expect of it in teaching and learning. Silver et al. 2013, 16, distinguishes between reflection and metacognition: "Reflection is . . . defined as a conscious exploration of one's own experiences . . . , and metacognition as the act of thinking about one's own thought processes." In educational settings, reflection and metacognition ask learners to integrate deliberate effort (e.g., training, practicing, trialing new or adjusted approaches) with conscious attention to learning, including the emotional aspects of that learning. We mention those emotional aspects because they inevitably arise; for example, learners need to be aware that learning and growth are change, which can be uncomfortable. Despite the potential for discomfort, learners need to remain open to and curious about experiences through reflection and metacognition. If they do, they will have found a powerful set of tools. For more about all this, see chapter 4.
20. See the chapter "Shitty First Drafts" in Lamott 1994. If you assign this to stu-

dents, though, be aware that Anne Lamott twice mentions suicide. Though she clearly means to be humorous, her rhetorical move of relating the despair of writing to the despair of suicide is not actually an exaggeration; no one should make light of just how desperate many developing writers actually feel.

21. We say "first step" because there are other strategies for discussing the writing of a paper. For example, students can be prompted to look for where and how the writer agrees or disagrees with previous research, how they make the case for the novelty or necessity of their work, or how they use and sequence narrative elements in the text. This practice makes visible the rhetorical "moves" or choices made by a writer (for a definition of "rhetorical move," see chapter 8, note 19). Gerald Graff and Cathy Birkenstein (2021, or earlier editions) provide a great framework for talking through rhetoric, with a fill-in-the-blank approach. Graff and Birkenstein's templates help students practice applying the rhetorical moves they see in model texts to their own writing, by making the end goal of wording visible.

22. Much of the soullessness and jargon in academic writing derives from the origins of Western science, when wealthy, White men worked to make research and scientific discourse an exclusive domain and then later "professionalized" science (Macfarlane 2007, A. Hull et al. 1982, Ideland 2018). At each stage, specific formalities in discourse were established to reinforce who did and did not have access (A. Hull et al. 1982, Chen et al. 2022, Singh 2022). One of the primary rhetorical strategies was conveying "objectivity" and "neutrality," with the parallel stance (intended or not) that only certain people are credible (Boucher et al. 2013, Wilder 2013, Treves 2019, Docot 2022). Bethann argues (Merkle 2024c) that it's a writing mentor's responsibility to point out this history to developing writers, as a key step in helping them find their own voices in science writing. But whatever the origins of our "academese," it has fed on itself as newer writers seek to imitate, or even outdo, the style of existing literature. On acronyms, see (and prepare to be astounded by) Barnett and Doubleday 2020. Plavén-Sigray et al. 2017 documents the increasing frequency of jargon in our literature, while Martínez and Mammola 2021 shows that papers with more jargon are cited less. Heard 2022c provides a more general discussion of these issues, and Heard 2022b and Merkle 2024c explore a common feedback loop: students and more experienced writers alike produce and reward writing with these features because that's what the scientific writing they're familiar with looks like.

23. Did you notice the contraction in that sentence? It's a fine example of the feedback loop in question. Many scientific writers are convinced that common English contractions have no place in scientific writing. There are no logical grounds for this (Heard 2015c; 2022c, 189). Instead, we perpetuate the notion that contractions don't belong in scientific writing simply because our past conviction means they aren't there as a model. As a result, we avoid them—despite the fact that their use can make writing more engaging and can give it broader tonal range and more nuanced emphasis (Silvia 2014). The feedback loop gets nicely summed up in *Moby Dick* (Heard 2020b), although not for writing style. Instead, Ishmael relates how navigators mistake breakers around the carcass of a dead whale for

breakers around a shoal, "and for years afterward, perhaps, ships shun the place" [because navigators remember avoiding the breakers but not why they did so] (Melville 1851, 344).

24. You were probably taught to use the passive; most of us were. Heard 2022c, 174–76, outlines the case for using the active voice in scientific writing. The history of this shifting convention (which is nicely traced by A. Gross et al. 2002) is interesting, and it's tied to the question of how our writing gains authority (Heard 2022c, 97–98). Briefly: early scientific writing was written primarily in the first-person active voice. However, the professionalization of science over the late 1800s and early 1900s was accompanied by an increasing emphasis on the (supposed) objectivity of science and scientists (Daston and Gallison 2007). Together, these forces led to Methods sections being simplified and stripped of personal detail. The passive voice became standard and reigned for a long time. We applaud the recent return to a preference for the active voice, which represents, in part, a realization that adopting the passive voice can only *pretend* objectivity, not achieve it. Similar debates, and similar changes through time, have played out in other writing genres, too; for instance, with respect to legal writing, see Coleman 1997 and Carpenter 2022.

25. The most obvious example is the arbitrary convention of declaring statistical significance at $P = 0.05$, rather than at $P = 0.04$, or 0.06, or 0.10, or rather than abandoning the line-in-the-sand view and interpreting the magnitude of P as a continuous measure of evidence (Heard 2015a). We teach cookbook tests (like cookbook grammar rules) in many other ways while shortchanging the logic behind them. Every student knows how to "do" a t-test; but few really understand what it is they're doing (Heard 2015d) and why, sometimes, they might not want to.

26. A. Gross, Harmon, and Reidy 2002 and Harmon and Gross 2007 are a good start. The former traces the history of writing style and conventions in the sciences, while the latter is something of a field guide crossed with an anthology that illustrates the conventions, shifts in them, and writing that rebels against them. Swales 1990 dissects many of our structure-and-content conventions, which is particularly useful for writers of Introductions and Discussions (where the conventions for content are less obvious than they are in Methods and Results). Heard 2022c summarizes many conventions of both content and style, and considers rationale when possible—for example, the extent of detail in the Methods (ibid., chap. 11) and the use of passive and active voice (ibid., chap. 18).

27. We'll talk about rhetoric a lot in this book, mainly to point you to research that is done in the field of rhetoric and composition. In essence, rhetoric is the way something is said, written, or otherwise conveyed (e.g., through visuals or music). One way of thinking about rhetoric that students find helpful is to recognize that its study asks whether and how communication efforts are effective. This involves considering the methods, mechanisms, timing, and way of calibrating to a reader, among other things. In other words, rhetoric goes beyond *what* is communicated to identifying *how* something is communicated (or at least, how the communication is attempted). There's a great overview of the concepts of

rhetoric and how important they are to communication in Losh et al. 2014, 35–65. The ways rhetoric tends to be used in science writing in particular are part of the discipline of science studies.

28. Lynne Truss's *Eats, Shoots, and Leaves* (2006) is an example: a grumpy book about perceived grammatical improprieties that became an unlikely bestseller. We will admit it's highly entertaining, but we're not recommending it for much else. Telling people they're wrong when you're not exactly right yourself isn't productive in any setting. Given the many hangups students have already acquired as a result of capricious grammar "rules," we argue that a mentor's time is better spent attending to students' needs—not what, in the end, amounts to style preferences.
29. The *Words at Play* blog and the *Usage Notes* resource from Merriam-Webster .com (https://www.merriam-webster.com/words-at-play/see-all, https://www .merriam-webster.com/grammar/usage-notes) frequently dissect supposed rules and errors, and are both erudite and entertaining—for example, in dissecting why "decimate" doesn't mean what the purists insist it does ("Regarding the Incorrect Use of 'Decimate,'" https://www.merriam-webster.com/grammar /the-original-definition-of-decimate). See also Fridland 2023.
30. *Finnegans Wake* (Joyce 1939) is a famous example of what happens when convention is ignored. Is it a great novel, Joyce's masterpiece, as Harold Bloom (1994) and others have claimed? It might be, but nobody really knows, because nobody understands it and even fewer have read it.
31. Beaufort 2007 describes these in more detail.
32. An emphasis on rhetoric in science writing *is* relevant and aligns with the deep history of writing instruction. Indeed, people have been using rhetoric to teach writing at least since Aristotle's classic *On Rhetoric*. Aristotle's impacts on how we do, think about, and write about science are profound. Interestingly, records indicate that his primary "textbook" was a collection of the notes his students took—thereby documenting what students gained from his teaching of rhetoric. For more on the classical history of rhetoric, see Bird et al. 2019b.
33. See Sanders-Reio et al. 2014 for a foundation in these ideas, and Inger Mewburn's extensive writings on the stress this causes students and strategies to avoid it (https://thesiswhisperer.com/).

CHAPTER TWO

1. Workplace needs were a central motivation of the Writing across the Curriculum movement (see chapter 4). Bethann has written about a longstanding gap between training and employer needs in wildlife biology (via a book review of *The Chicago Guide to Communicating Science*, Merkle 2019). Mewburn 2020 takes the argument further, asking why graduate students still write theses, and proposing alternatives that would build and demonstrate a more interesting range of competencies. Of course, these ideas are up against all the inertia of the academic training system (curricula, graduate admissions, competitive hiring, grant panels, publish-or-perish pressure on mentors, and so much more).
2. See Mewburn 2020 for a passionate argument to this effect.

3. We're not making a novel claim here. Higher-education pedagogy has long embraced transferable skills, although the definition has become quite vague (see, e.g., Johnson 2010, Olesen et al. 2021). We find Christof Nägele and Barbara Stalder's (2017, 739) definition useful: "Transferable skills are skills that are relevant and helpful across different situations and areas of life." Nägele and Stalder also point out that "such skills are often seen as a crucial factor adding to the employability of individuals. It is often assumed that transferable skills can be reused after transition to a new situation" (ibid.). The same skills are sometimes called employability, generic skills, job readiness, personal skills, or soft skills, and some people may speak of these as abilities, attributes, capacities, or competencies rather than skills. Kristoffer Olesen and coauthors (2021) grapple with defining transferable skills in a health-sciences context, with extensive review of literature. Kathleen Black Yancey and coauthors (2014) offer Teaching for Transfer, an entire curriculum and conceptual framework for teaching writing as a set of transferable skills.
4. For definitions of "reflection" and "metacognition," see chapter 1, note 19.
5. Heard 2022c, chap. 8–16, provides some history and examines these expectations at length.
6. If you assign op-eds, the Op-Ed Project is a great resource: https://www.theopedproject.org/communityresources.
7. Katalin Wargo (2020) provides an approachable review of authentic writing and the literature that substantiates its value, discussing a spectrum from latent to functional authenticity. According to her, "To have authentic purpose, the activity must serve the genuine purpose of communicating information to someone who wants or needs it" (539) and must actually involve the type of text that writers create outside the context of learning certain content or genres. For example, lab reports exist only in the context of undergraduate laboratory courses and thus are not an authentic writing activity, while drafting a methods section, entire manuscript, or magazine article on the topic of the same lab activity could be. See also Johns 2019 for discussion of how to help students understand the work of writing in a discipline (versus the typical mode of teaching students content and expecting them to somehow assimilate the rhetorical techniques needed to express it.)
8. Michele Eodice and coauthors (2016) used survey data to explore what made writing assignments "meaningful" to students. Almost 70% of respondents mentioned a writing assignment they found relevant to what they'd be writing in their future career (Eodice et al. 2016, 41). Other things that made writing assignments meaningful for students included agency in choosing topics, perception that the writing would be connected to past or future learning (especially on a topic they were passionate about), and awareness of progress in skills or learning. At least the first two of these are easier to achieve in authentic writing tasks than more rote ones.
9. Heard 2016 outlines the assignment and his experience as an instructor, and links to the student-written posts—which are still online and public-facing, seven years later as we write.

10. We've used "public" and "audience" as shorthand throughout the book. However, in the realms of science communication and public engagement, these terms are contested and increasingly avoided. More apt terms include "publics," "collaborators," "partners," and the like. But this terminology isn't our focus, and so we've opted for language that may be more familiar to most readers. For more, see Merkle et al. 2022 and Reed et al. 2024.
11. If you're unfamiliar with the phrase "issue-scoping report," you might prefer to think of it as a project proposal—akin to the more familiar thesis proposal, but specific to a particular project in a course. In such a report, students should specify the questions they'll ask in their project, the kinds of methods they'll use, the people who might be involved (collaborators, interviewees, stakeholders, etc.), and more. We call this an issue-scoping report because calling it a proposal misleads students into thinking they need to come up with answers even before they understand the questions they should ask and the people they're working to help, inform, or persuade.
12. Outside the undergraduate classroom, you might find the facilitation of authentic writing tasks less obvious. But you can extrapolate from the skills you'd want an undergraduate to develop to those you'd want to help grad students and postdocs develop—involving skill at tasks that are authentic and *not* exclusively academic. For example, some degree programs require PhD candidates to write a grant proposal rather than a more typical manuscript-style proposal. Other possibilities include preparing a policy brief, drafting a report to a funder or interested civic group, collaborating with local media on radio and/or print journalism about students' research, and so on. Many grad students are very interested in these kinds of writing opportunities even if it means doing work beyond the requirements of the degree program.
13. A great starting point is Merkle et al. 2022. Bethann also lectures, writes, and consults extensively on effective and ethical scicomm; see her professional website, www.commnatural.com, and her blog, *School of Good Trouble*, for relevant resources.
14. Broder et al. 2024.
15. For more on the deficit model (and why it generally doesn't work), see chapter 4 below (especially endnote 21) and Simis et al. 2016.
16. Kahneman 2013.
17. See Merkle et al. 2022 for more extensive discussion.
18. For extended discussions of the factors influencing public trust in science and individual scientists' credibility, see, for example, Sarewitz 2004, Scheufele 2013, Fiske and Dupree 2014, Donner 2014, Kotcher et al. 2017, and Iyengar and Massey 2018.
19. See, for example, Daston and Galison 2007, Ideland 2018, Treves 2019, and Broder et al. 2024.
20. Over a decade ago, Roger Bohn and John Short (2012) estimated that the average person in the United States was exposed to 34 gigabytes of information, including over 100,000 words, per day—*not even counting their exposure on the job*. (For comparison, this book, including its notes and tables, is a little under 100,000 words.) Imagine what that count must be up to now!

21. A typical adult in the United States spends only approximately 5% of their lifetime in a formal classroom; of that time, only a fraction is dedicated to science education (Falk and Dierking 2010).

CHAPTER THREE

1. Standard Academic English can be thought of as a "prestige dialect," and those who speak and write a form of English that differs from it can be disadvantaged. Useful perspectives on prestige dialects in academic writing and STEM communication include Rose 1985, G. Hull and Rose 1989, and Trachtenberg et al. 2018. However, note that in linguistics, the notion of a single, standard English is contested. The essential takeaway here is to recognize—and to help developing writers understand—that there are myriad variants of English. These variants play key roles in science writing and many other discourse spaces (e.g., Tremain 2023) and may influence the success of a writer's efforts to participate in these spaces.
2. Shaughnessy 1977, 85. This book, *Errors and Expectations*, was a watershed in writing instruction in higher education. In it, among other things, Mina Shaughnessy outlines the concept of "basic writer" as a specific, identifiable developmental level in postsecondary students. While some aspects of the book are now dated, it provides support for a host of "basic writer" challenges.
3. Treglia 2008.
4. For extensive discussion of this framework, see Elbow 1993.
5. See, for example, Kepner 1991 on error correction versus positive comments, with examples for additional-language learners. Agius and Wilkinson 2014 provides productive synthesis and recommendations. Note that there's potential for students to misunderstand positive comments that are actually hedging or softening the blow of critique rather than serving as genuine positive feedback (Hyland and Hyland 2001).
6. Agius and Wilkinson 2014.
7. Dressler et al. 2019 discusses types of feedback ranging from positive to negative, surface to actionable, etc. Their table 1 is especially helpful, as it provides examples of feedback from instructors categorized as surface-level, meaning-level, rhetorical, and so on. For example, they distinguish between general positive rhetorical feedback ("Great!" or "Good point!") and substantive positive rhetorical feedback ("Excellent section—clearly defines the purpose of the study.") The latter is, of course, more useable feedback.
8. For many reflection prompts and activities to reinforce the pleasurable aspects of writing, see Sword 2023.
9. See Kennell et al. 2017 for detailed discussion, and see the Purdue Online Writing Lab's succinct overview here: https://owl.purdue.edu/owl/general_writing/the_writing_process/feedback/instructor_guide_giving%20feedback.html.
10. We promise: this isn't an opinion—it's a body of knowledge that can transform how you mentor writers. We'll provide citations that establish that, but there are far too many (from decades of writing-instruction research) for us to cover exhaustively, especially in a single note! Meanwhile, we'll give you the takeaways from all that research. Taking advantage of the research to deliver effective, effi-

cient feedback requires acknowledging that some of our most central and intuitive feedback habits in scientific writing just don't work.

11. While some folks draw a distinction between a thesis and a dissertation, there's no universal agreement about what that distinction should be. We use "thesis" throughout to describe the research document submitted and defended by any graduate student at the end of their research training.
12. For more discussion about setting instructional goals (and gauging your progress toward them), see Merkle 2024a and the afterword.
13. Schwalm 1985 provides a good example of this "time limits = bad writing" reality. David Schwalm focused on academic-writing tasks and studied writing performance of students writing in English as an additional language. He reported that most students, when forced to write something new under a time constraint, "experienced partial or total linguistic collapse. . . . Grammatical, lexical, and syntactic skills that they *seemed* to have mastered disintegrated. The papers were nearly incomprehensible. . . . *Their skills developed in personal writing, especially sentence level skills, were not adequate to simple academic writing tasks*" (633, emphasis in the original).
14. We say "hope" because even with longer timeframes, you can expect most student writing to be first drafts unless you build in intermediate deadlines. Most students are juggling complex matrices of responsibilities and expectations, and your writing assignments are only one component. For myriad reasons, students prioritize other things over working ahead or incrementally on writing assignments. Granted, you'll sometimes see exceptions among students performing hypercompetence/perfectionism and students who deal proactively with personal hurdles (e.g., learning differences such as dyslexia). Susan Adams (1996) discusses these considerations, as well as the broader nuances of students with learning differences (while Adams focuses on writing in law school, there are many useful insights for the scientific-writing mentor). Of course, you'll see some students who can get by via last-minute drafting; but even their writing could get better with more deliberate, sequential tutelage.
15. For an intriguing rabbit hole, see K. Cho and MacArthur 2010 and Truscott and Hsu 2008 about the value of iterative writing processes for improving a single piece of writing versus sustained improvement of the writer's skills.
16. For concrete evidence that this approach works, we refer you to Bethann's students. She uses this approach in her classes, including the nearly twenty semester-long, writing-intensive undergrad and grad courses she has designed and taught. The approach also works in the English/composition courses Bethann has seen people trained in for a decade. As you might expect, most of the evidence supporting this process comes from the humanities. Some literature points to these principles at work in writing-intensive science courses, though that's a shorter stack of papers.
17. For more on this (and on writing feedback more generally), see Purdue Online Writing Lab, "Instructor's Guide for Giving Feedback": https://owl.purdue.edu/owl/general_writing/the_writing_process/feedback/instructor_guide_giving%20feedback.html.

18. For an ambitious examination of peer versus instructor feedback and a review of related research, see Anson and Anson 2017.
19. On signaling patterns of repeated errors, see Bean and Melzer 2021, 311.
20. An early presentation of the concept appears in Haswell 1983; for further exploration, see Bean and Melzer 2021, chap. 14.
21. Haswell 1983 argues that minimal marking might actually be applied for these other types of revision, though meaning needs to be conveyed clearly to the student. We're not completely on board. Using minimal marking for more-subtle issues risks sending the writer into the dark, guessing. While we might hope students would be spurred to think broadly and carefully about why we've marked a passage, we think this is too much to ask and suspect that mounting student frustration will just lead them to disengage with the task. We find that minimal marking is useful *in a specific context*; mentors should be careful about using it beyond those bounds (Treglia 2008, Dressler et al. 2019, Bean and Melzer 2021).
22. For more on the benefits of minimal marking, see McNeilly 2014 and Haswell 1983.
23. Haswell 1983, McNeilly 2014, and Lisman 1979 get into these ideas in useful detail. There's an important caveat, though: it's probably unfair to expect the same of students who are writing in English as an additional language (see chapter 8). Be alert to the possibility that minimal marking might send such students down fruitless rabbit-holes, and that you might need to provide more detailed feedback, either proactively or on request.
24. Schwalm 1985.
25. Schwalm 1985.
26. A targeted intervention Bethann was grateful to receive from an instructor when she was a sophomore in college. Steve, as you'd expect, has his own writing pecadillos. Or is it peccadillos?
27. Bartholomae 1980 (cited in Bean and Melzer 2021). Both of us regularly encourage students to do this.
28. This handout from thc Brandeis University Writing Program is an excellent starting point for reverse outlining: https://www.brandeis.edu/writing-program/resources/faculty/handouts/reverse-outlining.html. Your students might also appreciate an overview video from Melbourne University, "Editing: Reverse Outline" (https://youtu.be/XrtRkz15BFM?si=xUgpUjtwtL3Livdk), and Rachel Cayley's (2011) step-by-step instructions.
29. Tolman and Kremling 2016.
30. This, of course, is the whole premise of freshman composition and writing courses, though not all majors require them. Even when they do, disconnects between a campus composition program and your own science department may prevent these goals from being realized. We discuss possible remedies for these disconnects below, in chapters 5, 7, and 10.
31. Our six tips, of course, don't exhaust the list of ways you can make your feedback more efficient and effective. Bean and Melzer 2021, chap. 13–15, provides a more extensive discussion.

CHAPTER FOUR

1. For a more detailed history, see Townsend 2016. Bean and Melzer 2021, 19–21, provides a shorter summary and draws useful parallels between Writing across the Curriculum and the critical thinking movement.
2. Russell 2002, 12.
3. Adapted from Townsend 2016.
4. We recommend digging into the value of inter- and transdisciplinary training and research methods in STEM (and beyond). In addition to Bethann's own research on the topic, which includes Polfus et al. 2017 and Januchowski-Hartley et al. 2018, see also Carlson and Sullivan 1999, Aslan et al. 2014, and Grossman and Lockwood 2015.
5. See Bean and Melzer 2021 for a solid intro on this.
6. While you might not *think* you want to go so far as assigning poetry writing in your biochemistry course, you might be surprised by the range of writing genres that students will produce over their careers (even, maybe, poetry; see, e.g., Anand 2022, Conn 2023). There's also great pedagogical value to incorporating a broad range of genres in teaching (even, maybe, poetry). For example, see Januchowski-Hartley et al. 2018, a paper Bethann coauthored about poetry, creativity, and science. Bean and Melzer 2021, 46–56, also lists many possibilities and discusses the value of including multiple genres and explicitly discussing their different audiences and expectations.
7. You're getting used to our pointing to reflective writing assignments repeatedly in this book. There's a good reason for that: it's clear that asking students to reflect on their own writing (etc.) helps them think metacognitively (that is, think about their own thinking and learning). In the long run, this approach helps them improve as writers and to take more responsibility for that improvement. For a powerful argument in favor of student reflection, as well as citations of some key literature, see Bean and Melzer 2021, 231–37. For more on mentoring toward independent learning, see chapter 6.
8. Writing daily doesn't work for everyone (Sword 2016a, 2017). However, developing writers have rarely seen experienced writers doing the slogging effort of drafting-and-revision cycles that it takes to produce polished writing. Developing a consistent habit for writing is a valuable process for students; they just shouldn't feel like failures if they don't maintain a rigorous regime of writing every single day. It's also important to remember that writing involves many kinds of work, including searching for, reading, and notetaking on literature, drafting figures and tables, refining descriptions of methods, and much more; many students may not even recognize this work is part of writing. Introducing students to incremental, consistent writing, and how to schedule it, can be revelatory for them.
9. Merkle 2023b.
10. For useful discussions of writing-group formats, see Rockquemore 2010 and Silvia 2018, chap. 4.
11. Heard 2022c, 49–51.
12. Linda Adler-Kassner and Elizabeth Wardle (2015) first articulated these concepts,

but we reworded a couple to make them more accessible in a science writing environment. If these concepts feel a tad abstract, see page 10 of their follow-up book (Adler-Kassner and Wardle 2019) for a breakdown of the ideas that each concept comprises; the 2019 book also offers detailed discussion of how scholars and educators have found these concepts useful (or not).

13. Cameron, Nairn, and Higgins 2009 also provides a valuable perspective on emotions, knowledge, and academic identity.
14. Collecting these materials may initially require a bit of time commitment on your part. Alternatively, you could ask students (perhaps in small groups) to find readings across genres on a topic covered in the course. You'll likely want to do some quality control on the results, but that can itself be a learning and discussion opportunity.
15. We could go on at length about the many kinds of assignments you can incorporate in a WID course, how they might be assessed and weighted, and how they might be designed to scaffold toward increasingly sophisticated work. Instead, we'll point you to Bean and Melzer 2021, chap. 4 and 5, for a wide range of possibilities. You might also find useful Lisa Tremain's (2019) description of the course she teaches to introduce students to thinking like writers and the process of learning to write. While the course she describes isn't a WID course, the approach and even some of her suggested readings could be readily adapted to the first weeks of a WID course or embedded in a writing-focused course in your department. On the value of annotated bibliographies in science writing instruction, see Merkle 2022.
16. As a reminder, we distinguish between "scientific communication" and "science communication" (scicomm). The field of scicomm is concerned with activities typically recognized as public engagement, broader impacts, outreach, knowledge translation, and so on, while scientific communication is what is typically conducted in academic and technical settings (e.g., peer-reviewed papers, grant applications, conference presentations, and even teaching disciplinary courses). For further discussion of this distinction, see Merkle et al. 2022 and Broder et al. 2024.
17. On considering plain language and vocabulary that's shared, see Merkle et al. 2022 and A. N. B. Smith and Merkle 2021. For a review of how "knowledge translation" is applied in medical settings, see Gagliardi et al. 2016. There's considerable potential for this paradigm to be applied throughout scicomm and writing instruction. On shared values, see Merkle et al. 2022.
18. As Bethann and her coauthors discuss in Merkle et al. (2022), just giving people more information and expecting them to change their minds or behavior is known as the "deficit model." This approach is known to be ineffective and even counterproductive, particularly in contrast to other models of communication which seek to engage (e.g., consultation, dialogue) or even work with (e.g., collaboration, co-production) people who aren't typically science "partners" (e.g., communities and cultures with other ways of knowing). See also Simis et al. 2016.
19. Merkle et al. 2022.
20. See, for example, Broder et al. 2024.

21. Merkle et al. 2022.
22. For an extensive (though not exhaustive) intro to this literature, see Merkle et al. 2022.
23. See Bethann's syllabi for detailed examples of what this can look like: https://www.commnatural.com/syllabi.
24. If you're not actually responsible for a scicomm course, our advice is simple: encourage those you mentor to take scicomm courses in or beyond their degree program. Period. Any course is (probably) better than none.
25. For a step-by-step process that you could use in the classroom or beyond, see Merkle et al. 2022.
26. For context, see Broder et al. 2024.
27. For more discussion of how to identify and frame a maximum of three goals for your course, see Merkle 2024a.
28. For a definition of "transferable skills," see chapter 2, note 3.
29. Just as in any content course where you assign peer-reviewed literature, a scicomm course that brings in research must facilitate student reading and learning of that material. In Bethann's experience, students are both capable and appreciative of engaging with scicomm literature when such assignments are bolstered by the instructor's active support. Furthermore, students (and colleagues on up to administrators and funders) take tenets of scicomm such as "The deficit model doesn't work" more seriously when we reference, and they see, literature substantiating those claims. (For more on the deficit model, see note 18, above.) Using the scicomm literature meets the expectations of students and other scientists, who are accustomed to claims being substantiated with citations. Put another way, using scicomm literature to teach scicomm follows another central tenet of scicomm: "Calibrate to your audience."
30. COMPASS Message Box: https://www.compassscicomm.org/leadership-development/the-message-box/.
31. Here's what that means in Bethann's course: Students turn in low-stakes writing assignments and discussion posts every week. Over the course of the semester, they write at least 15,000 to 25,000 words (60–100 manuscript pages). That's a huge amount of writing, even for a capstone course in the three-course communication series required for undergraduates at Bethann's university. Some of that writing (a project scoping report, an annotated bibliography, and a public-facing blog post) undergoes numerous rounds of feedback and revision. Other assignments (e.g., scicomm tools reflections, choose-your-own-adventure assignments, questions for practitioner/scholar panels) are graded and receive feedback on the content/robustness of students' thinking (instead of being graded for the quality of the student's writing). All the major writing assignments are intensively scaffolded, and students work through them incrementally over several weeks. Thus, students practice several types of writing as well as learning different approaches and uses for drafting, revising, and polishing writing.
32. See Merkle 2023b.
33. For a concrete way of gauging how much work you're making for yourself and your students, you can use the Workload Estimator (https://cat.wfu.edu

/resources/workload2/) provided by the Center for the Advancement of Teaching at Wake Forest University. The calculator considers everything from discussion posts and word-count density of assigned readings to how complicated the concepts are and more.

34. See Merkle 2023b.
35. The University of Wisconsin–Madison (https://assessment.wisc.edu/student-learning-outcomes/writing-student-learning-outcomes/) and the American Association of Family Physicians (http://www.aafp.org/dam/AAFP/documents/cme/faculty_development/LearningObjectivesGuidelines.pdf) both offer useful overviews to help you write clear, measurable learning objectives.
36. A popular application of this approach is colloquially known as "ungrading." Ungrading methods have been touted as a powerful aspect of active pedagogy (see, e.g., Gorichanaz 2022, Crogman et al. 2023, and Talbert 2023). Although the term has been used to mean everything from portfolio grading to contract or standards-based grading, the original intent of ungrading is to provide only feedback, no grades, until the end of the semester, and only then to co-determine a course grade with the student. See Von Bergen 2023 for helpful discussion of all these approaches. Alternative assessment approaches can feel compelling or just trendy; we're sympathetic to both takes. We will caution, though, that many types of neurodiverse learners struggle mightily with some versions of ungrading; in our experience, most students benefit from specific, concrete, and sequential assignments with very explicit assessment schemes.
37. For further discussion of assessment approaches, see McGlynn 2020.
38. Some folks take rubrics for granted; others might have only heard of them within the last decade. A rubric is a document that clearly details expectations (e.g., content, sections of a document, number of references), with each expectation tied to performance indicators (e.g., meets, exceeds, does not meet expectations), and the grading associated with each performance level. Or, as Kiruthika Ragupathi and Adrian Lee (2020, 74) detail: "A rubric must have: evaluative criteria; quality definitions; and a scoring strategy." Ragupathi and Lee clarify that these three attributes, in combination, make rubrics pedagogical tools distinct from checklists or mere rating scales. They further note that rubrics can serve many purposes and be applied in a range of ways: analytic versus holistic; generic versus task-specific; teacher-centered versus student-centered. Their article is a valuable orientation to rubrics. For advice on building your own rubrics, see Reynders et al. 2020 for STEM-specific discussion; more generally, see Bean and Melzer 2021, chap. 12, and González-Chordá et al. 2016. If you've heard criticisms of rubric use, see Ernesto Panadero and Anders Jonsson's (2020) review, which finds little empirical evidence for those criticisms.
39. At Bethann's university, this is an increasingly common and mutually beneficial practice: science units get well-trained writing instructors, since many English/composition programs provide intensive pedagogical training to grad students while they teach (as instructors of record) in programs such as freshman composition. Partnering humanities departments get funded teaching assistantships (usually a scarce resource there) and can thus bolster their enrollment. The grad-

uate students doing this type of TA "exchange" gain experience instructing a wider range of courses.

40. For example, see the University of Wyoming's Learning Actively Mentoring Program (LAMP; https://www.uwyo.edu/science-initiative/lamp/index.html). LAMP is a comprehensive, sustained mentoring and professional development program with an emphasis on how to best adopt active learning strategies in high-enrollment courses. Bethann is currently partnering with LAMP and with collaborators at several other universities on a proposal to follow the LAMP model to specifically integrate writing and scicomm into science course instruction.

CHAPTER FIVE

1. A reminder, from chapter 3, that we prefer not to get hung up over distinctions between theses and dissertations; we use "theses" inclusively.
2. Perhaps you're surprised to hear us refer to peer reviewers as writing mentors—but good reviewers can provide valuable writing help. Peer review plays two functions (Heard 2019b): a gatekeeping function, recommending acceptance or rejection; and a manuscript-improvement function, helping make the manuscript better. Good reviewers understand that the latter is at least as important as the former.
3. We're going to provide some high-level suggestions here. For a detailed, step-by-step process complete with a facilitator's script, see Merkle 2023b. While that chapter discusses planning for a scicomm project, the same steps can be applied to any major writing project. Students may also appreciate the "magic to-do list" generator provided here: https://goblin.tools/.
4. Fisher et al. 2024.
5. This doesn't need to mean *every day* (see chapter 4); it just means doing something other than waiting for (and panicking at) the deadline.
6. Rewarding smaller successes is more sustainable. Newport 2016 argues that the average human's limit for generative work (like writing) is roughly four hours total per day. If developing writers are stalled out, it might be because they don't realize that three straight hours of writing is a lot.
7. For discussion of writing-group formats and how developing writers can best profit from them at various phases of a project, see Rockquemore 2010 and Silvia 2018, chap. 4. These materials and the associated process are required components of the Scholarly Writing Practices graduate program Bethann co-directs (Fisher et al. 2024).
8. See the appendix for recommended writing guides, and there especially Silvia 2018, Heard 2022c, Lamott 1994, and King 2020; all engage extensively with writer behavior.
9. Outlines come in many forms, and writers differ in which they find most helpful. Heard 2022c, 62–72, reviews many approaches to outlining, including word stacks, concept maps, and subhead and topic-sentence outlines.
10. Heard 2022c, 38–39, explores this idea, including the (hopefully obvious) point that data mockups and early Results text must always be clearly labeled as such, lest inattention lead someone to present the mockups as final results.

11. There's variation across fields and places as to whether a student writing a thesis is writing a scientific paper (or several) or a document that resembles a monograph. An increasingly common form is the "papers format" thesis, which is simply one or more papers sandwiched between a more traditionally styled general introduction and general discussion. Heard 2022c, 265–67, examines alternative thesis structures and the communicative functions they play.
12. For example, scientific writing can—believe it or not—include metaphors, beautifully stylish writing, and jokes (Heard 2014). It can be written less formally than is often the case, using short sentences, active voice, and even contractions (Heard 2022c, 174, 177, and 189, respectively). There's even some evidence that using humor in paper titles increases their impact on the field (Heard, Cull, and White 2023). Despite all this, and sadly, nearly everyone is advised at some point to make their scientific writing less interesting. If you (or your students) doubt our claim about the latitude you have, or need resources to bolster your argument for taking that latitude and running (writing) with it, see Steve's writing (and the occasional rant) about this on his blog and in Heard 2022c.
13. We admit to a special obsession with the scourge of acronyms. Their use in scientific writing has increased dramatically over the last century (A. Gross, Harmon, and Reidy 2002, Barnett and Doubleday 2020) and has reached hideous proportions. Adrian Barnett and Zoe Doubleday (2020) searched a corpus of about 18 million published abstracts in the health and medical sciences and found *over 1.1 million different acronyms*. That's (probably) more than the number of words in the English language (Michel et al. 2011)! Nearly a third of the acronyms appeared in a single paper's abstract but never again; plenty more appeared repeatedly but with different meanings. Using lots of acronyms will indeed make your scientific writing fit in with what's already in our literature. It's still a bad idea.
14. As (ahem) we do frequently in this book. Most writing rules are made to be broken, and breaking them judiciously can result in compelling text. Varying sentence length and structure, for example, is one crucial way that developing writers can explore stylistic expression even in technical writing—and varied sentences make for more pleasurable and sustainable reading.
15. Heard 2024b. And for evidence that there's little consensus on this point, see Fox 2024.
16. Tremain 2019, 2023.
17. For example, we're sure there will be writers who think the writing style we've adopted for this book is a bad choice. Here's why we're firm about our preferences: our goal is clear, engaging writing. This sets us up, for example, to prefer the active voice, to include a readable mix of simple sentences and short words and longer ones of both, to use humor, metaphors, and common contractions, to be more sparing with jargon and acronyms, etc. You may disagree; even if you don't, you will certainly work with other mentors who do. But if you think our scientific literature could be different and better, know that you aren't alone. These issues are explored at more length in Greene 2025 and Heard 2022c, among others.
18. By "colleague and coauthor" we mean coauthorship in the sense of two writers jointly and equally contributing to the writing task. This is not the same thing

as coauthorship in the sense of earned inclusion in a paper's list of authors. It's entirely possible for a coauthor (in the latter sense) to have done rather little authoring (in the former sense); witness, for example, the trend for papers to have hundreds or even thousands of latter-sense coauthors. (Aad et al. 2015, as of this writing, holds the record, with 5,154 coauthors—more coauthors than the paper has words.) For a fuller exploration of what list-of-authors coauthorship means and how to manage it, see Heard 2022c, 275–86.

CHAPTER SIX

1. Heard 2022b reflects on the tendency of developing writers to model existing literature, for better and often for much worse—and how the way we teach writing rewards this modeling.
2. For useful discussions of writing-group formats, see Rockquemore 2010 and Silvia 2018, chap. 4.
3. Merkle 2024d.
4. This notion that their writing is being "corrected," because they haven't written in the only right way, parallels other facets of the cognitive development of university students. It's common for students to arrive at university with a right/wrong view of knowledge, and to think of their education as the accumulation of the right answers (that is, of factually correct information). Bean and Melzer 2021, 21–23, 26–27, explores this and reviews relevant literature. Dweck's growth-mindset work (e.g., Dweck 2016) is an increasingly common framework for mentors and learners aiming to overcome the "fixed" mindset of right/wrong answers. Pajares 2003 reviews the role of self-efficacy beliefs in academic writing, showing how students' prior (especially negative) experiences with writing can drive them toward a right/wrong sense of writing.
5. You might wonder why we didn't say "learning to conduct a writing project independently." Well, that's important, too, but in most of the careers developing writers pursue, team writing is at least as important as solo writing. This is perhaps most familiar to us in the world of scientific papers. While early in the history of scientific publication (before about 1800) nearly all papers were solo authored (Beaver and Rosen 1978), since then, rates of coauthorship have been increasing across the sciences. Coauthorship rates reached around 90% in most fields by 2000 (Glänzel and Schubert 2004, Wuchty et al. 2007). The number of authors per paper has climbed as well (Wuchty et al. 2007). It's not just scientific papers, though; writing in almost any setting is increasingly likely to involve collaboration. That's why we say independence doesn't mean just the ability to write alone; it also means the ability to contribute to, or lead, team writing without ongoing supervision from a mentor.

CHAPTER SEVEN

1. Heard 2022c is the offending party here. Steve has written about proposal writing in a series of blog posts, though; see Heard 2022a for an entry point. If you're

looking for resources that treat proposal writing in more detail, try Friedland et al. 2018, Oruc 2011, or Schimel 2011.

2. See, for example, Higham 2020. For a list of recommended resources more generally, see the appendix.
3. Here are a few examples to get you started. Faculty Focus (https://www.facultyfocus.com/?s=writing) and Cult of Pedagogy (https://www.cultofpedagogy.com/?s=writing&submit=search) are resource sites that collect advice and tools from numerous contributors. Deanna Mascle, who runs the Morehead State University Writing Project, has a compilation of resources (https://metawriting.deannamascle.com/), as does Megan McIntyre, who directs the Writing Program at Sonoma State University (https://www.meganmmcintyre.com/resources). You'll find that exercises and other resources come in at least as many flavors as ice cream. Try the vanilla. Try the bubble gum or black licorice (Bethann's go-to). You'll certainly find takes on writing instruction that suit your goals and teaching style; just be sure to try a couple of flavors outside your usual, in case one becomes your new favorite.
4. On the TEA paragraph, see Merkle 2024b. Bethann worked as a journalist and editor for several years before teaching freshman composition. Despite all her experience as an editor, she struggled to articulate to students what the sequencing of a functional paragraph actually looks like and why. The TEA paragraph was the tool she needed to make coherent writing possible for most of her students.
5. We both post our course syllabi online. Steve has even taken the extra step to package up the exercises, assignments, and slide decks from his graduate-level scientific-writing course (Heard 2020c). Bethann provides syllabi for a variety of courses at https://www.commnatural.com/syllabi. Just remember that course development represents significant intellectual labor. Be as respectful in your use and citation of course materials as you would be of other intellectual property in your discipline.
6. This is not to say there aren't genuine norms and expectations in scientific writing that we do well to meet, and to teach. There are! But there are even more personal hobbyhorses and pet peeves. Steve tackles this issue extensively in *The Scientist's Guide to Writing* (Heard 2022c) and has vented a little more freely about SOURs (*s*trong *o*pinions *u*nmoored from *r*ationale) in Heard 2023b.
7. If you have students who are strong writers with an interest in supporting others' writing, this type of employment can provide them funding and valuable training. Students applying for these jobs needn't be steeped in literature or expert at writing in all genres. Writing centers are increasingly hiring discipline-specific peer consultants because they recognize that each discipline has particularities that discipline-familiar consultants are better equipped to address. You or your students can inquire at your local writing center to see (1) if STEM-familiar writing consultants are already on staff and (2) if it's possible for the writing center to hire some, or some more.
8. For an accessible essay you can draw from or make available to your students, see Rafoth 2010.
9. Fisher et al. 2024.

10. Practices like co-writing and other body-doubling strategies (wherein one person is working or doing a particular activity/task near you and thus supports or motivates you to do the same) may be especially helpful for neurodiverse people, such as those with ADHD (attention-deficit/hyperactivity disorder). See, for example, Hallowell and Ratey 2021, Anderson 2022, and preliminary results from Eagle et al. 2023.
11. A session with librarians is built into every iteration of the grad writing program Bethann co-leads, and of Steve's writing course, too. Students consistently remark that they were unaware of the range (and accessibility) of librarians' expertise.
12. For useful discussions of writing group formats, see Rockquemore 2010 and Silvia 2018, chap. 4. You can look online for advice on "critique" sessions to find practices (largely from creative writing and the studio arts) on how to facilitate critique/feedback groups. Graduate students we know have also self-organized feedback groups, with and without participation by more experienced writers such as postdocs and faculty. They report that such support can help students feel less isolated in their writing work.
13. As mentioned in note 11, in ADHD circles, the same setup is known as "body doubling" or "task doubling"; see Anderson 2022 and Hallowell and Ratey 2021.
14. See chapter 10 for detailed discussion of how you can structure this.
15. Whether and how this is possible will depend on graduate regulations at the students' university. Heard 2018 explores several issues and possibilities around co-authored thesis chapters.
16. See Anson and Anson 2017 and Brammer and Rees 2007 for accessible discussion of the history and application of research on course-based peer review. Bean and Melzer 2021, 243–51, also reviews the literature and has a helpful, practical take on using peer review in the classroom.
17. Instructors heading down this rabbit hole may also want to search on "peer response," "peer feedback," "peer revision," "group work," "writing groups," and the like.
18. Co-construction is "the joint creation of a form, interpretation, stance . . . or other culturally meaningful reality. The *co-* prefix in *co-construction* is intended to cover a range of interactional processes, including collaboration, cooperation, and coordination. However, co-construction does not necessarily entail affiliative or supportive interactions. An argument, for example, in which the parties express disagreement, is nonetheless co-constructed" (Jacoby and Ochs 1995, 171).
19. Again, see Anson and Anson 2017, Brammer and Rees 2007, and Bean and Meltzer 2021, all of which document real issues with peer review and provide evidence-based recommendations for overcoming those challenges.
20. On trust, see Brammer and Rees 2007.
21. Most of the literature on peer review comes from writing studies and from rhetoric and composition, and assumes that the goals of peer review are higher-level (organization, clarity, calibration to audience, writing as discovery, etc.). We're wholly on board with these goals (see chapter 3, especially tip 2). But we recog-

nize that there are lots of reasons to do peer review, and some are more mundane or involve later-project-stage goals, like line edits. To get the most out of the literature, just understand that most of it is working from a specific set of goals, and that you may need to adapt the advice if your goals differ.

22. We especially recommend Gannon 2020, McGlynn 2020, Lang 2021, and the classics by Paulo Freire (2018, 2021).
23. A similar approach can also be helpful in research labs. For example, you can ask graduate students to do tiered rounds of feedback before they submit drafts to you, such as (1) peer feedback, (2) postdoc or equivalent feedback, and (3) submission to the advisor. Similar iterative reviews can be useful at all career stages, of course.
24. For detailed discussion, we recommend starting with Bean and Melzer 2021, which provides extensive discussion of peer review along with exercises you can adopt. Anson and Anson 2017 and Brammer and Rees 2007 provide accessible discussions of the history and application of research on course-based peer review. DePeter 2020 and Kelly 2023 provide essays introducing undergraduates to doing peer review well. Finally, Côté 2018 specifically addresses peer-review facilitation for students who speak English as an additional language.
25. The types of resources and their availability on campus versus in a community setting varies by campus, community, state/province, and country. For example, in the US, many universities and colleges provide (basic) health care to students, while in Canada students receive basic health care through the provincial system, with the school they attend providing various kinds of supplemental coverage.
26. For example, PhD students with a documented condition of dyslexia may need (and must be offered, per federal disability protections in the US) extra time, someone to read them the prompts, and/or someone to facilitate basic corrections like spelling while the student is taking timed, written comprehensive exams.
27. Waldinger and Schulz 2023 provides a thorough and engaging synthesis of research on human happiness.

CHAPTER EIGHT

1. Ammon 2012. The 90% figure, though, is based simply on a count of papers, not on their readership or impact. Papers in languages other than English are less likely to be discovered by readers, in part because they're less likely to be included in indices such as Web of Science (Hanauer and Englander 2013, 7). They are thus, unsurprisingly, less likely to be read and cited (Whitehand 2005). They're also less likely to be given weight in academic assessments and, less formally, in researcher prestige (Corcoran et al. 2019, 4; Lillis and Curry 2010, 48ff; Waigandt et al. 2019; Muir and Solli 2019; Nauman 2019; Shahriari and Ghonsooly 2019). For these and other reasons, most scientists will want to publish in English whether or not they find that easy, and whether or not the need for them to do so is fair.
2. The construction "speaker/writer of English as an additional language" (EAL)

is a bit cumbersome, but it has replaced "non-native speaker" because it's less othering; and it's more accurate than "speaker/writer of English as a second language" (English often being a scientific writer's third or fourth language). In this chapter we'll connect you to some of the literature on EAL scientific writing, but it's useful if you know enough to find that literature yourself. The subject is generally called "English for research publication purposes," or a bit more broadly "English for academic purposes." Among the journals publishing work in the area are the *Journal of English for Specific Purposes*, the *Journal of English for Academic Purposes*, and the *Journal of Second Language Writing*.

3. For reviews of how EAL writers perceive their disadvantages, see Hanauer and Englander 2013 and Corcoran et. al. 2019. Unfortunately, literature assessing techniques for alleviating those disadvantages—that is, for mentoring EAL writers in scientific writing—is far less copious. That's a remarkable understatement, actually. It's easy to find publications describing particular interventions (workshops, writing center programs, etc.); but these are mostly descriptive and rarely provide evidence that one intervention works better than another. This chapter represents our attempt to provide the best guidance we can. Beyond that, Bird et al. 2019a offers some perspectives on fostering a writer's identity that you may also find helpful.
4. While this may seem too obvious to need literature support, it's a theme that crops up repeatedly in interviews with EAL writers. See, for example, S. Cho 2004.
5. We're cleverly ducking the issue of the extent to which these correlations reflect causation, and if so, in which direction. For a fascinating review of the interplay of linguistic background with cognitive processes including writing, reading, perception, reasoning, and memory, see Blasi et al. 2022. For a detailed exploration of how language origin influences English written texts, see Mur-Dueñas and Šinkūnienė 2018.
6. Amici et al. 2019.
7. See, for example, Gopen and Swan 1990. One can get too clever about this, though. While speakers of right-branching languages may be better at remembering the ends of lists than the beginnings, and vice versa for speakers of left-branching languages, *all* speakers are better still at remembering shorter lists! This is why we often suggest this simple revision to developing writers: split long sentences into multiple shorter sentences. This reduces the cognitive load placed on readers who need to hold information in their short-term memory until they can decode the sentence as a whole.
8. Although long lists have been provided for speakers of particular languages; see, for example, Hart 2017 for speakers of Chinese.
9. For a reasonably detailed treatment, see DiYanni and Hoy 2000, chap. 17. For a bewildering but fascinating thicket of advice, google "rules for use of articles in English."
10. See DiYanni and Hoy 2000, chap. 17.
11. Hinds 1987, Qi and Liu 2007; but for a quite different perspective, see MacKenzie 2015.

12. The study of such differences is called "contrastive rhetoric." We're offering generalizations; the ways that culture and language influence rhetorical approach are complex and also debated. For an introduction, see Kubota 1992 and Connor 1996. For a more recent review, see MacKenzie 2015.
13. Yes, we know: in previous lives we surely spoke one of these languages!
14. See, for example, Gosden 1996, Flowerdew 1999, and Englander 2009. For an interesting contrast, see Lorés-Sanz 2018, which finds that Spanish speakers from Spain are more deferential and thus reluctant to point to research gaps.
15. Such nuances of culture and language are not exclusive to EAL writers. Developing writers who grew up with English may have learned to read, write, and speak in settings that also make consistent production of standard academic English more difficult. As just one example, Bethann's writing and speech (in English) are permanently influenced by being a minority-language immigrant in French Canada. Even earlier, her conditioning to English writing included children's books from the 1950s, the "Old World" English of the King James Version of the Christian Bible, Dutch-inflected slang from her mother's family, and an appreciation for the Transcendental poets (Ralph Waldo Emerson and others). Her high school and university-level writing was thus loaded with flowery adjectives and complex and even convoluted sentence structures. She now recognizes similar impulses in students she mentors. Sometimes these are attempts to "sound smart" or academic; but sometimes they reflect a student's early experiences with language.
16. It's worth pointing out that nearly all these tips would be useful for English-as-first-language writers, too. We call attention to them here because they'll be particularly valuable in the EAL context. Similarly, EAL writers are, also, just regular writers, and they can benefit from the advice you'd give any student.
17. Krashen 1982 and Saville-Troike and Barto 2016 provide thorough reviews of the science of second-language acquisition, although neither is particularly focused on writing. Krashen 1982 was enormously influential but is somewhat dated now. Saville-Troike and Barto 2016 updates the science and provides explicit discussion of where Krashen 1982's arguments have held up and where they haven't, and how they influenced practices in second-language instruction.
18. Oliviera et al. 2014 provides a detailed protocol for a writer wishing to implement this strategy.
19. A "rhetorical move" is a bit of text that uses a particular linguistic structure and/or type of content to advance an argument in a particular way. Swales 1990 provides a thorough analysis of typical rhetorical moves in scientific writing; Heard 2022c outlines the way a near-universal set of moves occurs in scientific papers, easing the job of both writers and readers.
20. In the literature, such compilations are part of "corpus-based linguistics"—the empirical analysis of language patterns in genre- or discipline-specific writing (e.g., Gledhill 2000a, 2000b). There's good evidence that EAL writers appreciate and benefit from such tools (e.g., Lee and Swales 2006). Note the close parallel between such linguistic analysis and the "genre analysis" that describes common rhetorical devices or tactics in scientific writing (e.g., Swales 1990, 2004).

21. See the appendix for our recommendations of guidebooks and other resources more generally. Here we point to some that may be particularly useful for EAL writers. Online resources include Morley's Academic Phrasebank (https://www.phrasebank.manchester.ac.uk/), the Corpus of American Contemporary English (https://www.english-corpora.org/coca/), and the Google Ngram viewer (https://books.google.com/ngrams/). The first of these is specialized to academic writing, while the others are more general. Tagnin 2014 provides some advice about using online corpora. Guidebooks include Wallwork 2011 and Schuster et al. 2014. Graff and Birkenstein 2021 combines an accessible discussion of typical English-writing rhetorical strategies with extensive fill-in-the-blank models for rhetorical moves such as carving out a research space and stating a disagreement. Mewburn 2013 also provides several lists or cheat sheets specifically for academics; these detail uses of conjunctive adverbs and verbs for framing authorial voice, presenting specificity, etc.
22. SciPo-Farmácia: https://escritacientifica.sc.usp.br/scipo-farmacia/eng/index2.php; SWAN: https://cs.joensuu.fi/swan/. Aluísio and Feltrim 2014 discusses the use of these programs.
23. Ingley and Pack 2023 provides advice for making sure that the use of ChatGPT or equivalent serves the development of the writer, not just the polishing of the particular text at hand.
24. Gosden 1996 and Hanauer and Englander 2013, 48, discuss the incidence of translation among other strategies for EAL writers. Flowerdew 1999 and Lillis and Curry 2010, 94ff, among others, comment on the difficulties translators have working with scientific texts. Dissatisfaction with translated manuscripts is widely reported; for some examples, see Hanauer and Englander 2013, 48, 87, 96, and Lillis and Curry 2010, 95.
25. See, for example, Uzuner 2008. These helpers are often described in the literature as either "literacy brokers" (when the focus is more on writing in a particular genre) or "language brokers" (when the focus is more on general English text production) (e.g., Lillis and Curry 2010, 87ff).
26. These are especially common outside the English-speaking world (Corcoran et al. 2019); but English-speaking universities often have them, too—they may be labeled "English for research publication purposes" or something similar. EAL writers may also benefit from broader programs that have complementary goals. For example, Bethann co-founded and co-runs a two-semester Scholarly Writing Practices program, focused on building a healthy mindset and productive writing habits for graduate-level developing academic writers in all disciplines. Roughly half of the annual participants (approximately 50–75/year) are EAL speakers.
27. Levis and Levis 2003 describes both the difficulties of, and one pathway to success in, the very common model of courses taught by instructors with expertise in English language or composition, rather than the scientific disciplines.
28. The use of professional editors is, for example, very common in Nigeria (Omobowale et al. 2019) and Spain (Peréz-Llantada et al 2011). Ayokunle Omobowale and coauthors report that, unsurprisingly, heavy demand for these services means that unqualified individuals jump into the market; thus, the quality of the work is uneven.

29. Quantitative evidence is less abundant than one might think. However, Walker 2010 found nearly twice the incidence of plagiarism among international students for New Zealand business undergraduates, while Gilmore et al. 2010 found a similar but weaker pattern for EAL graduate students across disciplines in the United States. Both studies used plagiarism-detection software on large corpora of written assignments, avoiding problems with the more common analysis of self-reported plagiarism rates or rates of disciplinary proceedings (the latter could be more frequent for EAL writers either because of instructor bias or because their plagiarism is more easily noticed in contrast to their own text).
30. See, for example, Pennycook 1996, Pickering and Hornby 2005, Flowerdew 2008, Adiningrum and Kutielah 2011, and Bennett 2017. However, the literature is not monolithic on this point, with Maxwell et al. 2008, for instance, finding no such attitudinal differences between Asian and domestic students at two Australian universities.

CHAPTER NINE

1. For multiple takes on this fable from the Anglo-Saxon medieval period, see "The Changing Story of Cnut and the Waves," https://www.medievalists.net/2015/05/the-changing-story-of-cnut-and-the-waves/.
2. We especially like the Editors Canada approach for ethical editing of undergraduate or graduate texts (https://editors.ca/hire-an-editor/ethical-guidelines/). You can easily adapt their ethics statement and template agreement to your purposes.
3. StayFocusd: https://www.stayfocusd.com/; Leechblock NG: https://www.proginosko.com/leechblock/; Freedom: https://freedom.to.
4. Evernote: https://evernote.com; Capacities: https://capacities.io; Obsidian: https://obsidian.md. These are just three examples of many, and the entries in this field change frequently. As a result, before adopting any one system, a user should consider its ability to export content in a format that can be imported by others.
5. Asana: https://asana.com; Ayoa: https://www.ayoa.com/; Goblin Tools: https://goblin.tools/. There are of course many alternatives.
6. Scrivener: https://www.literatureandlatte.com/scrivener. As usual, there are many alternatives. Scrivener was developed with a particular focus on writing fiction and has tools for tracking characters and narrative arcs that may seem irrelevant to science writing, but some writers find it's worth experimenting with them anyway.
7. Dropbox: https://www.dropbox.com/paper/start. We assume few readers need help finding Word or Google Docs.
8. A related approach is to talk through writing ideas with someone unfamiliar with the content or the assignment. Students often find it helpful to discuss aloud what they plan to say—especially in early outlining/drafting phases, and at points when revision is needed but they're struggling to conceptualize what about the text really needs to change.
9. Look, we know you're thinking it, so we might as well admit that Steve (at least)

is old enough to remember when dictation was the domain of a human secretary clutching a steno pad and writing in inscrutable shorthand. This work was usually undervalued, as were those who did it. It often involved more than rote transcription, and it played a more important role than usually acknowledged in the development of the science we know today.

10. There is, unfortunately, an inequity element to the use of dictation software: available software works best for English, and in particular for "elite" dialects and accents such as BBC English ("received pronunciation") and midwestern "American." Other speakers will see reduced performance and may need to experiment to find the best among available apps. Fortunately, dictation software is improving rapidly in this context.
11. EndNote: https://endnote.com; Mendeley: https://www.mendeley.com; Zotero: https://www.zotero.org.
12. Remember our strong pitch in chapter 4 for the value of reflective assignments in general; see also Bean and Melzer 2021, 231–37.
13. Close et al. 2012, for example, reports both observational and experimental evidence that student achievement was not degraded—and may actually have improved—with the introduction of calculators to the classroom.
14. Roberts 2014 sketches the history of these and many other technological innovations in mathematics education, going back as far as the abacus. For writing instruction, Purcell et al. 2013 surveys US public (K–12) schoolteacher attitudes to, and experience with, digital writing tools, while C. Williams and Beam 2019 reviews literature on the impact of those tools on student writing skills and on challenges in the use of those tools. M. Li 2021 provides a detailed perspective on digital tools in EAL teaching (see also chapter 8).
15. Grammarly: https://www.grammarly.com/; Linguix: https://linguix.com/; Hemingway App: https://hemingwayapp.com/; Writer's Diet, https://writersdiet.com/.
16. Readable: https://app.readable.com/text/; Datayze: https://datayze.com/readability-analyzer; Lexile: https://hub.lexile.com/analyzer.
17. DuBay 2004 reviews readability scores in detail.
18. Bethann has students use readability analyzers explicitly during a lesson about jargon. The goal is to help students understand what the directive "Don't use jargon" actually means, and how avoiding jargon is a nuanced effort tied to the calibration of a text to its ideal readers. Specifically, she prompts students to take a text and use the readability calculators to drop *and* increase the readability score by three levels. Doing so helps students see how adjusting word choice or sentence structure actually changes readability for an anticipated reader.
19. Developing writers do, unfortunately, transgress—whether it's the occasional sentence plagiarized from Wikipedia or the submission of an entire essay or thesis purchased from an online essay mill. Parnther 2020 reviews data on the incidence of plagiarism and other academic dishonesty; while there's great variation, it's clear that neither undergraduate- nor graduate-student plagiarism is rare. What you do about this connects to the much broader topic of academic dishonesty in general. A full discussion is beyond the scope of this book, but we

encourage you to think about this carefully. A good policy will be enforceable, transparent, and respectful of the honesty of most students. The widespread use of plagiarism checkers is controversial in part because it seems not to satisfy the latter two conditions (and also sometimes flags honest writing as plagiarism). McGlynn 2020 considers academic dishonesty in pragmatic detail, with a focus on undergraduate programs, while Jamieson and Howard 2019 reviews and considers academic dishonesty from historical and philosophical perspectives. Mbutho and Hutchings 2021 and Parnther 2020 provide useful discussions of plagiarism with a global perspective, including student knowledge of, and attitudes toward, various forms of plagiarism. Finally, most universities have their own policies (with which mentors should be familiar), and university libraries are often good sources of advice for both mentors and students. See chapter 8 for some further discussion of plagiarism in the particular context of writers for whom English is an additional language.

20. The Editors Canada resource we've already mentioned (see note 2) includes a detailed permission form that a student and thesis supervisor can complete, clarifying exactly what kind of editing is permitted in a given context.

21. Well, OK, that isn't quite the *only* question. While on-campus writing services may be free, professional editing costs money. Thus, if you allow or even encourage students to have their work edited, you should think about whether all your students are able to access that service, and how you might be able to provide equitable access. Of course, this isn't an issue unique to editing—tuition, campus accommodation, and textbooks present similar equitability issues (Murphy and Shelley 2020). We admit that our book and your policies can't solve all these problems.

22. One problem in any treatment of LLMs is that because of their recent appearance and rapid development, there's an enormous amount written about them, and yet most of it is either breathless (whether hype or prophecy of doom) or hyperspecific to narrow details of the latest model. So, while we'll point you to sources in this section, we acknowledge that for LLMs, there isn't yet the broad and useful kind of literature that exists for writing pedagogy. Three recent books, however, may be worth reading. Mark Carrigan's *Generative AI for Academics* (2024) is perhaps the most bullish on the technology, outlining ways for academics to use LLMs in their work—including, but not limited to, their writing. We especially appreciate Carrigan's (2024, 8) argument for "thinking with generative AI" rather than "using generative AI as a substitute for thought." Arvind Narayanan and Sayash Kapoor's *AI Snake Oil* (2024) covers more forms of so-called "AI" than just LLMs, but their chap. 4 is LLM-focused and includes a good review of the harms and dangers of LLM use. Their concern, in general, is in separating uses of "AI" that in their view do not (and cannot ever) work from those that do work and can be harnessed for productive ends. They place LLMs squarely in the latter category, even while recognizing both pros and cons. Finally, John Warner's *More Than Words: How to Think About Writing in the Age of AI* (2025) is the most resolutely anti-LLM. Despite recognizing a few productive ways to use LLMs to help with tasks like reading and summarizing documents, Warner argues at length that almost any involvement of LLMs with writing devalues both the pro-

cess and the product while impeding writer development. "I think," he writes, "there's something, many things, essential about what happens as humans write that cannot and should not be outsourced to automation posing as intelligence" (52). He does not seem to believe it possible for writers to (in Carrigan's words) think with LLMs rather than using them instead of thinking—and it would certainly be worrisome if he is right (but see Lira et al. 2025). These three books as a set provide a good deep dive into LLMs and the cases for and against their use, if you're interested in going that far.

23. If you need more than this very brief explanation, there's an early but excellent introduction to ChatGPT in an *Ezra Klein Show* podcast with Gary Marcus: https://www.nytimes.com/2023/01/06/opinion/ezra-klein-podcast-gary-marcus.html. Wolfram 2023 and Stöffelbauer 2023 go into more technical depth.
24. Marcus 2025 provides a good explanation of why LLMs hallucinate. Attempts to make LLMs that *do* know whether the things they say are true have so far involved combining the text prediction of an LLM with web search to retrieve (hopefully) accurate information. This is called "retrieval augmented generation" (RAG-LLM; Martineau 2023). It's not yet clear how well this will work, and it may work best for very simple applications where queries are limited to the contents of just a few documents—for example when a chatbot acts as an interface to a user manual or set of HR policies. Indeed, it's not even yet clear how we can measure how well a RAG-LLM works (see, e.g., Salemi and Zalani 2024). An alternative approach, "reinforcement learning," involves the model using trial and error and comparing solutions to different queries; a ChatGPT model featuring this approach was released in September 2024. So far, at least, this strategy hasn't had much success and in fact may actually have made the problem worse (OpenAI 2025).
25. See, for example, Bethann's reflection (Merkle 2023a) on the citations and statistics ChatGPT invented when she prompted it to draft a scicomm grant proposal. Aksenfeld 2025 also provides a startling example of the fake-citation problem.
26. While LLMs are too new to have accumulated a substantial literature documenting their missteps, we expect that to change rapidly. Meanwhile, Hussam Alkaissi and Samy McFarlane (2023) and Jerome Goddard (2023) are among those to have documented ChatGPT hallucinations. Ken Masters (2023) discusses the detection of ChatGPT hallucinations in a manuscript submitted for journal publication. In the legal world, Molly Bohannon (2023) reported on the consequences for a lawyer who allowed ChatGPT-hallucinated citations to appear in a court filing. Emilio Ferrara (2023), Nicole Gross (2023), and Valentin Hofman and coauthors (2024) review the issue of ingrained biases in ChatGPT and other LLMs, and Dhruv Agarwal, Mor Naaman, and Aditya Vashistha (2025) provide evidence of bias toward Western styles and cultural references.
27. Rillig et al. 2023 provides a brief review of environmental impacts of LLMs, both positive and negative, including indirect environmental impacts that touch on social ethics, too. See more in the next note.
28. These figures are drawn from Tomlinson et al. 2024, Chen 2025, Dauner and Socher 2025, and Huggingface.co 2025. The range reflects in part decisions about

which energy usage to include and which not to. This isn't simple—for instance, some estimates include energy used in model training while some don't; some also include the energy cost of manufacturing the chips that run the model. None as far as we know include the energy cost of feeding the engineers who designed them or of how those engineers got to work. It also reflects different assumptions about the fossil-fuel intensity of energy generation (moving LLM servers to places where carbon intensity is lower can provide large reductions; Chien et al. 2023). Finally, different LLMs are more or less computationally intensive but also more or less computationally *efficient* (Tomlinson et al. 2024). ChatGPT, which is one of the most widely used models, is unfortunately one of the least efficient. Overall, it seems safe to think of per-query carbon footprints on the order of a gram or two of CO_2, with plenty of potential for reduction through model choice or gains in efficiency. For a deeper dive and careful comparisons of LLM writing with *other* human activities, see Masley 2025a, 2025b and Ritchie 2025. The streaming video estimate is from data summarized in Schien et al. 2024, although such estimates are complex and hotly contested (see, e.g., Kamiya 2020 and Marks and Przedpełski 2022). The per-kilometer driving cost estimate is from the US Environmental Protection Agency (US EPA 2024) and would be lower in countries with more efficient vehicle fleets. Estimates for cheeseburgers vary widely (as do cheeseburgers) but are shocking: for instance, 1.9 kg CO_2/burger (Babakhani et al. 2019) or 3.6 kg CO_2/burger (Petrat-Melin and Dam 2023). We expect estimates like this to change rapidly, for several reasons. Some future LLMs will be more complex, using more energy; others will be more efficiently coded, using less (the latest versions of DeepSeek, in particular, seem to be considerably less energy-intensive than their rivals, although estimates of how much "considerably" means vary wildly; O'Donnell 2025, Turner 2025). Finally, the energy efficiency of computation and of data centers, in general, is increasing rapidly.

29. Tomlinson et al. 2024. Where an LLM is used to assist a writer rather than to replace one, the same kind of gains are possible. Noy and Zhang 2023 found a 40% decrease in time required for college-educated workers to complete writing tasks when they had access to ChatGPT. If they shut off their laptops after completing the task (granted, a questionable assumption!), the energy saved would easily exceed that consumed by their ChatGPT queries.
30. Li et al. 2025.
31. For the US, recent guidance from the United States Copyright Office (2025b) suggests that AI training without permission will likely be illegal; however, court cases, not such guidance, will be needed to settle the issue. W. Cho 2023 reviews the status, as of 2023, of multiple lawsuits alleging copyright infringement in LLM training and output. In two recent US cases (Salvaggio 2025), the judges found against plaintiffs' claims of illegal training, but in an interesting twist, each judge explained that the legal reasoning behind their decision was narrow and that other plaintiffs bringing similar claims were likely to prevail. One party supporting these lawsuits, and authors' rights more generally, is the Authors Guild, a US-based professional organization (Knight 2023).

32. Copyright law might prohibit the use of some writing in this way, although there are complexities around fair use and educational use and although copyright law varies from country to country. In addition, a library's contract with a publisher may prohibit uploading material to an LLM. At least some material with a Creative Commons license may be appropriate for use in this way (Walsh 2023), or (of course) material in the public domain or unpublished material you've written yourself. Because these issues are complex and rapidly changing, it's best to look for advice from your institution's copyright office or its library. Meanwhile, some general advice on copyright, privacy, and LLMs is available from Portland University's Clark Library: https://libguides.up.edu/ai/copyright.
33. See Perrigo 2023; we're not aware of peer-reviewed literature.
34. We say "considering multiple positions" rather than "taking and arguing for one position" because we have both soured on the value of debate structures in teaching. Taking and arguing for a position tends to cement that position in the arguer, without necessarily being very effective at persuading listeners. (This is something practitioners of scicomm should remember as well.) For more on this, see Merkle 2025.
35. For more discussion, see Heard 2023a.
36. It's the prerogative of a journal or other publishing outlet to set requirements on submissions, even if some authors might resent those requirements. *Nature*, for example, requires authors to disclose any use of LLMs (in either the Methods or the Acknowledgments). This policy is accompanied with some interesting discussion in a recent editorial (*Nature*, 2023). Zielinski et al. 2023 discusses responses and requirements of journals more generally with respect to LLM use by authors. One interesting reason for journal objections to LLM use might matter to some authors, too: depending on the legal jurisdiction, an author may not hold the copyright to text produced using an LLM. For example, in the United States, the direct product of an LLM is not copyrightable, although text produced by a human with LLM assistance may be, depending on the balance of human/LLM contributions (United States Copyright Office 2025a).
37. If you're not sure you believe that LLM use is widespread, see Kobak et al. 2025 for data on the occurrence of LLM-influenced text in the biomedical literature. Tseng and Warschauer 2023 makes the argument about teaching effective and ethical use at more length and suggests a general framework for building LLMs into teaching for those who choose to do so.
38. McGlynn 2020 (11–13, 33–35, and especially 150ff) provides general recommendations for handling issues of enforcement and academic misconduct. While none of it is specific to LLM use, you will easily be able to extrapolate.
39. Ingley and Pack 2023 is a short paper that develops this critical idea, connecting it to more general evidence-based recommendations from second-language acquisition. Wang et al. 2024 considers postsecondary students' self-reported success at using LLMs to build writing skills (and review previous studies doing similar work). Carrigan 2024 makes an extended argument that's a bit more general, emphasizing that one should be "thinking with generative AI" rather than "using generative AI as a substitute for thought" (8). For a thoughtful but much more

pessimistic assessment of this possibility, see Warner 2025; but for empirical evidence siding more with Carrigan than Warner, see Lira et al. 2025.

40. For more on using LLMs as a reviewer (among other things), see Mewburn 2024, 2025.
41. Fisher et al. 2024 and Wang et al. 2024 provide data on these student attitudes about feedback.
42. Ingley and Pack 2023 discusses, among other things, the importance of student reflection and/or discussion in active learning about LLM assistance.
43. Our list of suggestions is based in part on our own teaching experience, but also draws on a number of sources that you can consult for more detail, different takes, and alternatives. Among these is an edited volume by Anette Vee and coeditors (2023) that provides a large collection of LLM-involved teaching activities. These come with instructor reflections but not with direct evidence of effectiveness. Further activities and discussion are available in Ingley and Pack 2023 and G. R. Smith et al. 2024, and from the Weitz Center for Creativity, Carleton College: https://www.carleton.edu/writing/resources-for-faculty/working-with-ai/incorporating-ai-tools/.
44. See note 32.
45. DuBay 2004 reviews readability scores in detail.
46. For an example of this kind of exercise in action, albeit with a different set of authors, see Heard 2024a.
47. For the advisability of using debate structures, see note 34.
48. Suggestions, examples, and discussion are available from the Weitz Center for Creativity, Carleton College: https://www.carleton.edu/writing/resources-for-faculty/working-with-ai/. A course or lab policy will likely be informed by broader institutional policies or guidelines, if your institution has them. For examples, see Dingemanse 2024 and University of Tartu (n.d.).
49. For a more thorough discussion of the use of LLMs in assessment, including both practicality and ethics, see an exchange among Zhai et al. 2022, T. Li et al. 2023, and Zhai and Nehm 2023. Steiss et al. 2024 provides some empirical assessment of LLM assessment quality.
50. We are not the only ones who are uneasy. John Warner (2025, 239–41) is even more so, and provides a spirited argument that LLMs should never be involved in assessment.
51. Jacob Steiss and coauthors (2024) examined LLM assessments in comparison with human assessments of the same texts and found that the LLM was generally not as good—except when it came to assessing a text against the assigned rubric, where it (slightly) *outperformed* humans. They suggest that given inevitable tradeoffs of assessment quality versus time, there may be worthwhile use cases for integrating LLMs into assessment workflows.
52. See Heard 2025 for an example, which was worked in some detail and includes discussion of writing flaws that ChatGPT in particular easily incorporated and other flaws (like spelling errors) that it struggled to incorporate.
53. As of the time of writing, available tools include Schemely (https://schemely.app/), MagicSchool (https://www.magicschool.ai/), and Eduaide (https://www

.eduaide.ai), but we expect particular tools to come and go; a web search on "ai lesson generator" should show you a more recent selection. These tools are often targeted at K-12 education—likely because that's a bigger market—but can be applied at other levels, too.

54. The incorporation of digital tools (and new writing forms arising from digital media) into writing programs is widespread, and digital literacy is widely considered to be an important goal when writing is taught (Sheffield 2016).

CHAPTER TEN

1. What we're pointing to is the generally held notion (at least in North America) that children should learn to read and write in a happy, supportive "story time" environment. But the instructional model committed to fun (or at least pleasurable and self-expressive) writing shifts in middle or high school to an emphasis on hard, good-at-it-or-not, quasi-academic writing. It should be no surprise that this stark change in the framing of writing has negative impacts on many a developing writer's sense of self-efficacy, identity as a writer, and motivation to write at all. For more on this front, see Pajares 2003, a review of the literature on the role of self-efficacy in writing.
2. For more detailed discussion, see Mazak 2019.
3. Bethann has one retired professor friend who spent most of his career at an R-1 university in the US Mountain West. There, a sequence of four writing-intensive content courses were required for the undergraduate STEM program he taught in. However, he reports (despairingly) that while those courses are still required, his colleagues no longer emphasize writing in them.
4. We're aware that there are as many ways to complete a graduate proposal as there are programs granting graduate degrees. However, as we note in table 10.1, we find the proposal-as-funding-application model especially appealing because it does what all the evidence points to: makes a major writing task feel more authentic and relevant, less like a massive busywork assignment. We also like the model more common in parts of Canada and Europe, wherein a graduate student does their proposal and comprehensive exam in their first year, rather than what is more typical in the US, where students do the proposal and exams after they've conducted most of their research. See again: relevance versus busywork.
5. Steve's Canadian upbringing has brought him to a muddle of British and American usages that from a Canadian perspective is perfectly normal, but that from any other perspective is a hot mess. For those interested in the (trivial) issue at hand: conventions for the placement of punctuation around quotation marks differ, with the American system simpler but the British more logically consistent. But it's complicated; for MLA style notes on this point see https://style.mla.org/the-placement-of-a-comma-or-period-after-a-quotation/, or for a longer read, see the University of Sussex guide to punctuating quotations at https://www.sussex.ac.uk/informatics/punctuation/quotes/marks.
6. Our framework is necessarily limited by room on the page and your attention, so it doesn't have every nuance we'd include if someone made us the (mostly) be-

nevolent supreme rulers of an entire curriculum. We tried to provide enough detail for you to envision a student's progression as they gain skills and confidence in writing, but not so much as to risk myriad "yeah, but" thoughts as you consider all the complexities and constraints of your own program.

7. Indeed, in most STEM degrees we're aware of in Canada, undergraduate students do not take a freshman composition course. These students may not receive *any* dedicated writing instruction in their entire four-year degree (Steve didn't).
8. For example, the required writing/communications course sequence at the University of Wyoming sets a strict cap of 24 students per section. In a department like Bethann's, with an average graduating cohort of 125 to 140 undergraduate students, it's simply not possible for faculty to offer enough sections of even one such course, let alone a sequence of three to four. And yet most science departments we're aware of would prefer to teach undergraduates through their own writing-focused courses, if only it seemed logistically possible.
9. See chapter 7 regarding the opportunities and challenges of reaching out to experts from other areas of your campus.
10. Your colleagues may argue that every journal has its own distinct style guide. They're correct, of course. In an ideal world, those of us publishing in the peer-reviewed literature would band together and call for a single, universal style for all our journals. We'd save ourselves (to say nothing of editors, peer reviewers, and copyeditors) so much time spent fiddling with minutiae like whether to put a comma between the name and date of an in-text citation! While that paradise is distant, a common curricular style guide can pay off even while you teach developing writers that there are stylistic preferences that vary by individual and by publication.
11. For a definition of "scaffolding," see the introduction, note 4; see also further discussion in chapters 1 and 4.
12. For a starting point on this theme, see Walvoord 2010. We'll note that this approach intersects with the much-debated Bloom's taxonomy. This isn't the time or place to get into that; suffice it to say that there are varying takes on how useful or accurate that model is (see, e.g., Booker 2007, Stanny 2016, Talbert 2017, and Larsen et al. 2022). In any case, many frameworks can help you articulate and nurture students' conceptual development via writing instruction and beyond (see, e.g., Irvine 2017, 2021).

AFTERWORD

1. Heard 2024c.
2. DeCosta and Roen 2015 provides a wonderful discussion of how your mentorship can be assessed to help you improve at it, not just to judge whether you should stay employed. It maps those five dimensions of scholarship onto six recognized criteria for assessing scholarship: clear goals, adequate preparation, appropriate methods, significant results, effective presentation, and reflective critique. The companion table for assessing teaching and learning is also helpful.

3. See, for example, Darrin Murray 2019 and Marissa Bell and Lewis 2022.
4. The SMART goal concept was introduced in Doran 1981, but since then the acronym has proven quite flexible. It's most often interpreted as specifying goals that are *s*pecific, *m*easurable, *a*chievable, *r*elevant, and *t*ime-bound. EASY goals are *e*nergizing, tied to your *a*gency, *s*mall, and truly *y*ours (Berdahl 2022).
5. See, for starters, the University of Wisconsin–Madison (https://assessment.wisc.edu/student-learning-outcomes/writing-student-learning-outcomes/) and the American Association of Family Physicians (http://www.aafp.org/dam/AAFP/documents/cme/faculty_development/LearningObjectivesGuidelines.pdf), along with DeCosta and Roen 2015 and Bean and Melzer 2021.
6. For a good start, Dayton 2015 and Bean and Melzer 2021 are valuable resources for thinking about how to measure, and achieve, student success in writing. More broadly, Walvoord 2010 provides a good introduction to measuring student success in postsecondary education. As a starting point, keep in mind that good grades aren't the goal—improved writers are.
7. On a "just in time" approach, see Merkle 2024a.
8. On work from and not for affirmation, see Montgomery 2019.

References

Aad, George, and Atlas Collaboration, CMS Collaboration. 2015. "Combined Measurement of the Higgs Boson Mass in *Pp* Collisions at $\sqrt{s}$=7 and 8 TeV with the ATLAS and CMS Experiments." *Physical Review Letters* 114 (19): 191803.

Adams, Sophie, Sheree Bekker, Yanan Fan, Tess Gordon, Laura J. Shepherd, Eve Slavich, and David Waters. 2022. "Gender Bias in Student Evaluations of Teaching: 'Punish[ing] Those Who Fail to Do Their Gender Right.'" *Higher Education* 83 (4): 787–807.

Adams, Susan Johanne. 1996. "Because They're Otherwise Qualified: Accommodating Learning Disabled Law Student Writers." *Journal of Legal Education* 46 (2): 189–215.

Adiningrum, Tatum Syarifah, and Salah Kutielah. 2011. "How Different Are We? Understanding and Managing Plagiarism between East and West." *Journal of Academic Language and Learning* 5 (2): A88–98.

Adler-Kassner, Linda, and Elizabeth Wardle, eds. 2015. *Naming What We Know: Threshold Concepts of Writing Studies*. Logan: Utah State University Press.

Adler-Kassner, Linda, and Elizabeth Wardle, eds. 2019. *(Re)Considering What We Know: Learning Thresholds in Writing, Composition, Rhetoric, and Literacy*. Logan: Utah State University Press.

Agarwal, Dhruv, Mor Naaman, and Aditya Vashistha. 2025. "AI Suggestions Homogenize Writing Toward Western Styles and Diminish Cultural Nuances." Proceedings of the 2025 CHI Conference on Human Factors in Computing Systems (New York, NY, USA), CHI '25, April 25, 1–21.

Agius, Natalie M., and Ann Wilkinson. 2014. "Students' and Teachers' Views of Written Feedback at Undergraduate Level: A Literature Review." *Nurse Education Today* 34 (4): 552–59. https://doi.org/10.1016/j.nedt.2013.07.005.

Aksenfeld, Rita. 2025. "Springer Nature Book on Machine Learning Is Full of Made-up Citations." *Retraction Watch*, June 30. https://retractionwatch.com/2025/06/30/springer-nature-book-on-machine-learning-is-full-of-made-up-citations/.

Alexander, Elizabeth S., and Anthony J. Onwuegbuzie. 2007. "Academic Procrastination and the Role of Hope as a Coping Strategy." *Personality and Individual Differences* 42 (7): 1301–10. https://doi.org/10.1016/j.paid.2006.10.008.

Alkaissi, Hussam, and Samy I. McFarlane. 2023. "Artificial Hallucinations in ChatGPT: Implications in Scientific Writing." *Cureus* 5 (2): e35179. https://doi.org/doi:10.7759/cureus.35179.

Aluísio, Sandra M., and Valéria D. Feltrim. 2014. "Writing Tools." In *Writing Scientific Papers in English Successfully*, edited by Ethel Schuster, Haim Levkowitz, and Osvaldo N. Oliviera Jr., 115–50. Andover, MA: Hypertek.com.

Amici, Federica, Alex Sánchez-Amaro, Carla Sebastián-Enesco, Trix Cacchione, Matthias Allritz, Juan Salazar-Bonet, and Federico Rossano. 2019. "The Word Order of Languages Predicts Native Speakers' Working Memory." *Scientific Reports* 9:1124.

Ammon, Ulrich. 2012. "Linguistic Inequality and Its Effects on Participation in Scientific Discourse and on Global Knowledge Accumulation—with a Closer Look at the Problems of the Second-Rank Language Communities." *Applied Linguistics Review* 3 (2): 333–55. https://doi.org/10.1515/applirev-2012-0016.

Anand, Madhur. 2022. *Parasitic Oscillations: Poems*. Toronto: McClelland & Stewart.

Anderson, Linda. 2022. "The Body Double: A Unique Tool for Getting Things Done." *ADDA—Attention Deficit Disorder Association* (blog), October 24, 2022. https://add.org/the-body-double/.

Anson, Chris M. 2015. "Technology and Transparency: Sharing and Reflecting on the Evaluation of Teaching." In *Assessing the Teaching of Writing: Twenty-First Century Trends and Technologies*, edited by Amy E. Dayton, 99–117. Logan: Utah State University Press.

Anson, Ian G., and Chris M. Anson. 2017. "Assessing Peer and Instructor Response to Writing: A Corpus Analysis from an Expert Survey." *Assessing Writing* 33 (July): 12–24. https://doi.org/10.1016/j.asw.2017.03.001.

Aslan, Clare E., Malin L. Pinsky, Maureen E. Ryan, Sara Souther, and Kimberly A. Terrell. 2014. "Cultivating Creativity in Conservation Science." *Conservation Biology* 28 (2): 345–53. https://doi.org/10.1111/cobi.12173.

Babakhani, Nazila, Andy Lee, and Sara Dolnicar. 2020. "Carbon Labels on Restaurant Menus: Do People Pay Attention to Them?" *Journal of Sustainable Tourism* 28 (1): 51–68. https://doi.org/10.1080/09669582.2019.1670187.

Barnett, Adrian, and Zoe Doubleday. 2020. "The Growth of Acronyms in the Scientific Literature." *eLife* 9 (July): e60080. https://doi.org/10.7554/eLife.60080.

Bartholomae, David. 1980. "The Study of Error." *College Composition and Communication* 31 (3): 253–69. https://doi.org/10.2307/356486.

Bean, John C., and Dan Melzer. 2021. *Engaging Ideas: The Professor's Guide to Integrating Writing, Critical Thinking, and Active Learning in the Classroom*. 3rd ed. Hoboken, NJ: Jossey-Bass.

Beaufort, Anne. 2007. *College Writing and Beyond: A New Framework for University Writing Instruction*. 1st ed. Logan: Utah State University Press.

Beaver, D. deB, and R. Rosen. 1978. "Studies in Scientific Collaboration: I. The Professional Origins of Scientific Co-authorship." *Scientometrics* 1 (1): 65–84. https://doi.org/10.1007/BF02016840.

Belcher, Wendy L. 2019. *Writing Your Journal Article in Twelve Weeks: A Guide to Academic Publishing Success*. 2nd ed. Chicago: University of Chicago Press.

Bell, Marissa, and Neil Lewis. 2023. "Universities Claim to Value Community-Engaged Scholarship: So Why Do They Discourage It?" *Public Understanding of Science* 32 (3): 304–21. https://doi.org/10.1177/09636625221118779.

Bell, Myrtle P., Daphne Berry, Joy Leopold, and Stella Nkomo. 2021. "Making Black Lives Matter in Academia: A Black Feminist Call for Collective Action against Anti-Blackness in the Academy." *Gender, Work & Organization* 28 (S1): 39–57. https://doi.org/10.1111/gwa0.12555.

Belland, Brian R. 2014. "Scaffolding: Definition, Current Debates, and Future Directions." In *Handbook of Research on Educational Communications and Technology*, 4th ed., edited by J. Michael Spector, M. David Merrill, Jan Elen, and M. J. Bishop, 505–18. New York: Springer.

Bennett, Karen. 2017. "The Geopolitics of Academic Plagiarism." In *Publishing Research in English as an Additional Language: Practices, Pathways and Potentials*, edited by Margaret Cargill and Sally Burgess, 209–20. Adelaide, Australia: Adelaide University Press.

Benson, Beth Kemp. 1997. "Coming to Terms: Scaffolding." *English Journal* 86 (7): 126–27. https://doi.org/10.2307/819879.

Berdahl, Loleen. 2022. "How to Set Goals That Aren't Completely Delusional." Substack newsletter. *Academia Made Easier* (blog), January 3, 2022. https://loleen.substack.com/p/how-to-set-goals-that-arent-completely.

Bird, Barbara, Doug Downs, I. Moriah McCracken, and Jan Rieman, eds. 2019a. *Next Steps: New Directions for/in Writing about Writing*. Logan: Utah State University Press.

Bird, Barbara, Doug Downs, I. Moriah McCracken, and Jan Rieman. 2019b. "Writing about Writing: A History." In *Next Steps: New Directions for/in Writing about Writing*, edited by Barbara Bird, Doug Downs, I. Moriah McCracken, and Jan Rieman, 13–20. Logan: Utah State University Press.

Blasi, Damián E., Joseph Henrich, Evangelia Adamou, David Kemmerer, and Asifa Majid. 2022. "Over-reliance on English Hinders Cognitive Science." *Trends in Cognitive Sciences* 26 (12): 1153–70. https://doi.org/10.1016/j.tics.2022.09.015.

Bloom, Harold. 1994. *The Western Canon: The Books and School of the Ages*. New York: Harcourt Brace and Company.

Boblett, Nancy. 2012. "Scaffolding: Defining the Metaphor." *Studies in Applied Linguistics and TESOL* 12 (2): 1–16. https://doi.org/10.7916/salt.v12i2.1357.

Bohannon, Molly. 2023. "Judge Fines Two Lawyers for Using Fake Cases from ChatGPT." Forbes. 2023. https://www.forbes.com/sites/mollybohannon/2023/06/22/judge-fines-two-lawyers-for-using-fake-cases-from-chatgpt/.

Bohn, Roger, and James Short. 2012. "Measuring Consumer Information." *International Journal of Communication* 6:980–1000.

Bong, Way Kiat, and Weiqin Chen. 2021. "Increasing Faculty's Competence in Digital Accessibility for Inclusive Education: A Systematic Literature Review." *International Journal of Inclusive Education*, 1–17. https://doi.org/10.1080/13603116.2021.1937344.

Booker, Michael J. 2007. "A Roof without Walls: Benjamin Bloom's Taxonomy and the Misdirection of American Education." *Academic Questions* 20 (4): 347–55.

Boucher, Cheryl J., Georgina S. Hammock, Selina D. McLaughlin, and Kelsey N. Henry. 2013. "Perceptions of Competency as a Function of Accent." *Psi Chi Journal of Psychological Research* 18 (1): 27–32.

Brammer, Charlotte, and Mary Rees. 2007. "Peer Review from the Students' Perspective: Invaluable or Invalid?" *Composition Studies* 35 (2): 71–85.

British Ecological Society. 2013. *A Guide to Peer Review in Ecology and Evolution*. London: British Ecological Society.

Broder, E. D., Bethann G. Merkle, M. Balgopal, E. Weigel, S. Murphy, J. J. Caffrey, E. Hebets, et al. 2024. "Use Your Power for Good: An Applied Framework for Overcoming Institutional Injustices Impeding SciComm in the Academy." *BioScience* 74 (11): 747–69. https://doi.org/10.1093/biosci/biae080.

Cameron, Jenny, Karen Nairn, and Jane Higgins. 2009. "Demystifying Academic Writing: Reflections on Emotions, Know-How and Academic Identity." *Journal of Geography in Higher Education* 33 (2): 269–84.

Carlson, Lawrence E., and Jaquelyn F. Sullivan. 1999. "Hands-On Engineering: Learning by Doing in the Integrated Teaching and Learning Program." *International Journal of Engineering Education* 15 (1): 20–31.

Carpenter, Jacob M. 2022. "The Problems, and Positives, of Passives: Exploring Why Controlling Passive Voice and Nominalizations Is about More Than Preference and Style." *Legal Communications and Rhetoric: JALWD*, Marquette Law School Legal Studies Paper No. 07, 19.

Carrigan, Mark. 2024. *Generative AI for Academics*. Thousand Oaks, CA: SAGE Publications.

Cataldi, Emily F., Christopher T. Bennett, and Xianglei Chen. 2018. "First-Generation Students: College Access, Persistence, and Postbachelor's Outcomes." NCES 2018421. National Center for Education Statistics. https://nces.ed.gov/pubsearch/pubsinfo.asp?pubid=2018421.

Cayley, Rachel. 2011. "Reverse Outlines." *Explorations of Style* (blog), February 9, 2011. https://explorationsofstyle.com/2011/02/09/reverse-outlines/.

Chen, Christine Yifeng, Sara S. Kahanamoku, Aradhna Tripati, Rosanna A. Alegado, Vernon R. Morris, Karen Andrade, and Justin Hosbey. 2022. "Systemic Racial Disparities in Funding Rates at the National Science Foundation." *eLife* 11 (November): e83071. https://doi.org/10.7554/eLife.83071.

Chen, Sophia. 2025. "How Much Energy Will AI Really Consume? The Good, the Bad and the Unknown." *Nature* 639 (8053): 22–24. https://doi.org/10.1038/d41586-025-00616-z.

Chien, Andrew A., Liuzixuan Lin, Hai Nguyen, Varsha Rao, Tristan Sharma, and Rajini Wijayawardana. 2023. "Reducing the Carbon Impact of Generative AI Inference (Today and in 2035)." In *Proceedings of the 2nd Workshop on Sustainable Computer Systems*, 1–7. Boston: ACM. https://doi.org/10.1145/3604930.3605705.

Cho, Kwangsu, and Charles MacArthur. 2010. "Student Revision with Peer and Expert Reviewing." In "Unravelling Peer Assessment," special issue, *Learning and Instruction*, 20 (4): 328–38. https://doi.org/10.1016/j.learninstruc.2009.08.006.

Cho, Seonhee. 2004. "Challenges of Entering Discourse Communities through Publishing in English: Perspectives of Nonnative-Speaking Doctoral Students in the United States of America." *Journal of Language, Identity & Education* 3 (1): 47–72.

Cho, Winston. 2023. "Authors Sue OpenAI Claiming Mass Copyright Infringement of Novels." *Hollywood Reporter*, June 29, 2023. https://www.hollywoodreporter.com/business/business-news/authors-sue-openai-novels-1235526462/.

Close, Sean, Elizabeth Oldham, Gerry Shiel, Therese Dooley, and Michael O'Leary. 2012. "Effects of Calculators on Mathematics Achievement and Attitudes of Ninth-Grade Students." *Journal of Educational Research* 105 (6): 377–90. https://doi.org/10.1080/00220671.2011.629857.

Coleman, Brady. 1997. "In Defense of the Passive Voice in Legal Writing." *Journal of Technical Writing and Communication* 27 (2): 191–203. https://doi.org/10.2190/HN2D-AVL9-7XK4-EVVC.

Conn, Jan. 2023. *Peony Vertigo*. Kingston, ON: Brick Books.

Connor, Ulla M. 1996. *Contrastive Rhetoric: Cross-Cultural Aspects of Second Language Writing*. Cambridge, UK: Cambridge University Press.

Cooke, S. J., N. Young, M. R. Donaldson, E. A. Nyboer, D. G. Roche, C. L. Madliger, R. J. Lennox, J. M. Chapman, Z. Faulkes, and J. R. Bennett. 2021. "Ten Strategies for Avoiding and Overcoming Authorship Conflicts in Academic Publishing." *FACETS* 6 (January): 1753–70. https://doi.org/10.1139/facets-2021-0103.

Corcoran, James N., Karen Englander, and Laura-Mihaela Muresan, eds. 2019. *Pedagogies and Policies for Publishing Research in English: Local Initiatives Supporting International Scholars*. Milton Park, UK: Routledge.

Côté, Robert A. 2018. "Teaching Writing Students How to Become Competent Peer Reviewers." *English Teaching Forum* 56 (4): 16–23.

Crogman, Horace T., Kwame O. Eshun, Maury Jackson, Maryam A. TrebeauCrogman, Eugene Joseph, Laurelle C. Warner, and Daniel B. Erenso. 2023. "Ungrading: The Case for Abandoning Institutionalized Assessment Protocols and Improving Pedagogical Strategies." *Education Sciences* 13 (11): 1091. https://doi.org/10.3390/educsci13111091.

Daston, Lorraine, and Peter Galison. 2007. *Objectivity*. New York: Zone Books.

Dauner, Maximilian, and Gudrun Socher. 2025. "Energy Costs of Communicating with AI." *Frontiers in Communication* 10 (June). https://doi.org/10.3389/fcomm.2025.1572947.

Dayton, Amy E., ed. 2015. *Assessing the Teaching of Writing: Twenty-First Century Trends and Technologies*. Logan: Utah State University Press.

Deanna, Rocío, Bethann Garramon Merkle, Kwok Pan Chun, Deborah Navarro-Rosenblatt, Ivan Baxter, Nora Oleas, Alejandro Bortolus, et al. 2022. "Community Voices: The Importance of Diverse Networks in Academic Mentoring." *Nature Communications* 13 (1): 1681. https://doi.org/10.1038/s41467-022-28667-0.

de Brey, Cristobal, Lauren Musu, Joel McFarland, Sidney Wilkinson-Flicker, Melissa Diliberti, Anlan Zhang, Claire Branstetter, and Xiaolei Wang. 2019. "Status and Trends in the Education of Racial and Ethnic Groups 2018." NCES 2019038. https://nces.ed.gov/pubsearch/pubsinfo.asp?pubid=2019038.

DeCosta, Meredith, and Duane Roen. 2015. "Assessing the Teaching of Writing: A

Scholarly Approach." In *Assessing the Teaching of Writing: Twenty-First Century Trends and Technologies*, edited by Amy E. Dayton, 13–30. Logan: Utah State University Press.

DePeter, Ron. 2020. "How to Write Meaningful Peer Response Praise." In *Writing Spaces*, edited by Dana L. Driscoll, Mary Stewart, and Matthew A. Vetter, 3:40–51. Anderson, SC: Parlor Press.

Dicks, Matthew. 2018. *Storyworthy: Engage, Teach, Persuade, and Change Your Life through the Power of Storytelling*. Novato, CA: New World Library.

Dingemanse, Mark. 2024. *Generative AI and Research Integrity*. Radboud University Nijmegen. https://doi.org/10.31219/osf.io/2c48n.

DiYanni, Robert, and Pat C. Hoy. 2000. *The Scribner Handbook for Writers*. 3rd ed. Boston: Allyn & Bacon.

Docot, Dada. 2022. "Dispirited Away: The Peer Review Process." *PoLAR: The Political and Legal Anthropology Review* 45:124–28.

Donner, Simon D. 2014. "Finding Your Place on the Science—Advocacy Continuum: An Editorial Essay." *Climatic Change* 124:1–8.

Doran, George T. 1981. "There's a SMART Way to Write Management's Goals and Objectives." *Journal of Management Review* 70:35–36.

Dressler, Roswita, Man-Wai Chu, Katie Crossman, and Brianna Hilman. 2019. "Quantity and Quality of Uptake: Examining Surface and Meaning-Level Feedback Provided by Peers and an Instructor in a Graduate Research Course." *Assessing Writing* 39 (January): 14–24. https://doi.org/10.1016/j.asw.2018.11.001.

DuBay, William H. 2004. *The Principles of Readability*. Costa Mesa, CA: Impact Information. Available at https://files.eric.ed.gov/fulltext/ED490073.pdf.

Duffy, Meghan A. 2017. "Last and Corresponding Authorship Practices in Ecology." *Ecology and Evolution* 7 (21): 8876–87. https://doi.org/10.1002/ece3.3435.

Dweck, Carol. 2016. "What Having a 'Growth Mindset' Actually Means." *Harvard Business Review*, January 13, 2016.

Eagle, Tessa, Leya Breanna Baltaxe-Admony, and Kathryn E. Ringland. 2023. "Proposing Body Doubling as a Continuum of Space/Time and Mutuality: An Investigation with Neurodivergent Participants." In *Proceedings of the 25th International ACM SIGACCESS Conference on Computers and Accessibility*, 1–4. ASSETS '23. New York: Association for Computing Machinery. https://doi.org/10.1145/3597638.3614486.

Elbow, Peter. 1993. "Ranking, Evaluating, and Liking: Sorting Out Three Forms of Judgment." *College English* 55 (2): 187–206. https://doi.org/10.2307/378503.

Englander, Karen. 2009. "Transformation of the Identities of Nonnative English-Speaking Scientists as a Consequence of the Social Construction of Revision." *Journal of Language, Identity & Education* 8 (1): 35–53. https://doi.org/10.1080/15348450802619979.

Eodice, Michele, Neal Lerner, and Anne Ellen Geller. 2016. *The Meaningful Writing Project*. Logan: Utah State University Press.

Falk, John H., and Lynn D. Dierking. 2010. "The 95 Percent Solution." *American Scientist* 98 (6): 486–93.

Ferrara, Emilio. 2023. "Should ChatGPT Be Biased? Challenges and Risks of Bias in Large Language Models." Available at arXiv, https://doi.org/10.48550/arXiv.2304.03738.

Fisher, Rick. 2024. "Reading like a Writer: A Workshop Activity." *School of Good Trouble* (blog), March 20, 2024. https://www.commnatural.com/post/reading-like-a-writer.

Fisher, Rick, Makayla Kocher, Joshua Clapp, and Bethann Garramon Merkle. 2024. "Meaningful Results with Limited Resources: Evidence from a Program to Support Graduate Students' Scholarly Writing." *College Teaching*, early view. https://doi.org/10.1080/87567555.2024.2407121.

Fiske, Susan T., and Cydney Dupree. 2014. "Gaining Trust as Well as Respect in Communicating to Motivated Audiences about Science Topics." *Proceedings of the National Academy of Sciences* 111: 13593–97. https://doi.org/10.1073/pnas.1317505111.

Flowerdew, John. 1999. "Writing for Scholarly Publication in English: The Case of Hong Kong." *Journal of Second Language Writing* 8 (2): 123–45. https://doi.org/10.1016/S1060-3743(99)80125-8.

———. 2008. "The Non-Anglophone Scholar on the Periphery of Scholarly Publication." *AILA Review* 20 (1): 14–27. https://doi.org/10.1075/aila.20.04flo.

Fowler, H. Ramsey, Jane E. Aaron, and Michael Greer. 2022. *The Little, Brown Handbook*. 14th ed. New York: Pearson.

Fox, Jeremy. 2024. "Poll Results: Should the Introduction Section of a Scientific Paper End with a Statement of the Main Results?" *Dynamic Ecology* (blog), March 12, 2024. https://dynamicecology.wordpress.com/2024/03/12/poll-results-should-the-introduction-section-of-a-scientific-paper-end-with-a-statement-of-the-main-results/.

Fox, Jeremy, and Bethann G. Merkle. 2019. "Teaching Science Writing in University Courses (Poll Results and Commentary)." *Dynamic Ecology* (blog), January 9, 2019. https://dynamicecology.wordpress.com/2019/01/09/teaching-science-writing-in-university-courses-poll-results-and-commentary/.

Freire, Paulo. 2018. *Pedagogy of the Oppressed: 50th Anniversary Edition*. 4th ed. New York: Bloomsbury Academic.

———. 2021. *Pedagogy of Hope: Reliving Pedagogy of the Oppressed*. New York: Bloomsbury Academic.

Fridland, Valerie. 2023. *Like, Literally, Dude: Arguing for the Good in Bad English*. New York: Viking.

Friedland, Andrew J., Carol L. Folt, and Jennifer L. Mercer. 2018. *Writing Successful Science Proposals*. 3rd ed. New Haven, CT: Yale University Press.

Gagliardi, Anna R., Whitney Berta, Anita Kothari, Jennifer Boyko, and Robin Urquhart. 2016. "Integrated Knowledge Translation (IKT) in Health Care: A Scoping Review." *Implementation Science* 11 (1): 38. https://doi.org/10.1186/s13012-016-0399-1.

Gannon, Kevin M. 2020. *Radical Hope: A Teaching Manifesto*. 1st ed. Morgantown: West Virginia University Press.

García-Peña, Lorgia. 2022. *Community as Rebellion: A Syllabus for Surviving Academia as a Woman of Color*. Chicago: Haymarket Books.

Giasson, Mischa. 2016. "The Cobblestone Tiger Beetle." Entomological Society of Canada, January 29, 2016. https://esc-sec.ca/2016/01/29/the-cobblestone-tiger-beetle/

Gilmore, Joanna, Denise Strickland, Briana Timmerman, Michelle Maher, and David Feldon. 2010. "Weeds in the Flower Garden: An Exploration of Plagiarism in Graduate Students' Research Proposals and Its Connection to Enculturation, ESL, and Contextual Factors." *International Journal for Educational Integrity* 6 (1). https://doi.org/10.21913/IJEI.v6i1.673.

Glänzel, Wolfgang, and András Schubert. 2004. "Analysing Scientific Networks through Co-Authorship." In *Handbook of Quantitative Science and Technology Research: The Use of Publication and Patent Statistics in Studies of S&T Systems*, edited by F. H. Moed, Wolfgang Glänzel, and Ulrich Schmoch, 256–76. Dordrecht: Kluwer Academic Publishers.

Glasman-Deal, Hilary. 2020. *Science Research Writing: For Native and Non-Native Speakers of English*. 2nd ed. Singapore: World Scientific Publishing.

Gledhill, Christopher J. 2000a. *Collocations in Science Writing*. Tübingen, Germany: Gunter Narr Verlag.

———. 2000b. "The Discourse Function of Collocation in Research Article Introductions." *English for Specific Purposes* 19 (2): 115–35.

Goddard, Jerome. 2023. "Hallucinations in ChatGPT: A Cautionary Tale for Biomedical Researchers." *American Journal of Medicine* 15 (2): e35179. https://doi.org/10.1016/j.amjmed.2023.06.012.

González-Chordá, Víctor M., Desirée Mena-Tudela, Pablo Salas-Medina, Agueda Cervera-Gasch, Isabel Orts-Cortés, and Loreto Maciá-Soler. 2016. "Assessment of Bachelor's Theses in a Nursing Degree with a Rubrics System: Development and Validation Study." *Nurse Education Today* 37 (February): 103–7. https://doi.org/10.1016/j.nedt.2015.11.033.

Gopen, George, and Judith Swan. 1990. "The Science of Scientific Writing." *American Scientist* 78:550–58.

Gorichanaz, Tim. 2022. "'It Made Me Feel like It Was Okay to Be Wrong': Student Experiences with Ungrading." *Active Learning in Higher Education*, May, 14697874221093640. https://doi.org/10.1177/14697874221093640.

Gosden, Hugh. 1996. "Verbal Reports of Japanese Novices' Research Writing Practices in English." *Journal of Second Language Writing* 5 (2): 109–28. https://doi.org/10.1016/S1060-3743(96)90021-1.

Graff, Gerald, and Cathy Birkenstein. 2021. *They Say / I Say*. 5th ed. New York: W. W. Norton.

Greene, Anne E. 2025. *Writing Science in Plain English*. 2nd ed. Chicago: University of Chicago Press.

Grogan, Kathleen E. 2020. "Writing Science: What Makes Scientific Writing Hard and How to Make it Easier." *Bulletin of the Ecological Society of America* 102 (1): e01800. https://doi.org/10.1002/bes2.1800.

Gross, Alan G., Joseph E. Harmon, and Michael Reidy. 2002. *Communicating Science: The Scientific Article from the 17th Century to the Present*. New York: Oxford University Press.

Gross, Nicole. 2023. "What ChatGPT Tells Us about Gender: A Cautionary Tale about Performativity and Gender Biases in AI." *Social Sciences* 12 (8): 435. https://doi.org/10.3390/socsci12080435.

Grossman, Ali, and Jeff Lockwood, dirs. 2015. *The UCross Experiment: A Cross-Pollination of Arts and Sciences.* http://www.youtube.com/watch?v=a051ANDxIDA.

Hallowell, Edward M., and John J. Ratey. 2021. *ADHD 2.0: New Science and Essential Strategies for Thriving with Distraction—from Childhood through Adulthood.* New York: Ballantine Books.

Hanauer, David Ian, and Karen Englander. 2013. *Scientific Writing in a Second Language.* Anderson, SC: Parlor Press.

Harmon, Joseph E., and Alan G. Gross. 2007. *The Scientific Literature: A Guided Tour.* 1st ed. Chicago: University of Chicago Press.

Hart, Steve. 2017. *English Exposed: Common Mistakes Made by Chinese Speakers.* Hong Kong: Hong Kong University Press.

Haswell, Richard H. 1983. "Minimal Marking." *College English* 45 (6): 600–604. https://doi.org/10.2307/377147.

Haynes, Chayla, Leonard Taylor, Steve D. Mobley Jr., and Jasmine Haywood. 2020. "Existing and Resisting: The Pedagogical Realities of Black, Critical Men and Women Faculty." *Journal of Higher Education* 91 (5): 698–721. https://doi.org/10.1080/00221546.2020.1731263.

Heard, Stephen B. 2014. "On Whimsy, Jokes, and Beauty: Can Scientific Writing Be Enjoyed?" *Ideas in Ecology and Evolution* 7 (1): 64–72.

———. 2015a "Is 'Nearly Significant' Ridiculous?" *Scientist Sees Squirrel* (blog), November 16, 2015. https://scientistseessquirrel.wordpress.com/2015/11/16/is-nearly-significant-ridiculous/.

———. 2015b. "Was Barbara Cartland a Genius? And Are You?" *Scientist Sees Squirrel* (blog), August 11, 2015. https://scientistseessquirrel.wordpress.com/2015/08/11/was-barbara-cartland-a-genius-and-are-you/.

———. 2015c. "Why Do Not We Use Contractions in Scientific Writing?" *Scientist Sees Squirrel* (blog), October 1, 2015. https://scientistseessquirrel.wordpress.com/2015/10/01/why-do-not-we-use-contractions-in-scientific-writing/.

———. 2015d. "Why Do We Make Statistics So Hard for Our Students?" *Scientist Sees Squirrel* (blog), October 6, 2015. https://scientistseessquirrel.wordpress.com/2015/10/06/why-do-we-make-statistics-so-hard-for-our-students/.

———. 2016. "Student Blogging on Insect Conservation: A Success Story." *Scientist Sees Squirrel* (blog), January 29, 2016. https://scientistseessquirrel.wordpress.com/2016/01/29/student-blogging-on-insect-conservation-a-success-story/.

———. 2018. "Can a Thesis Chapter Be Coauthored?" *Scientist Sees Squirrel* (blog), April 10, 2018. https://scientistseessquirrel.wordpress.com/2018/04/10/can-a-thesis-chapter-be-coauthored/.

———. 2019a. "Reclaiming Voice in Scientific Writing." *Scientist Sees Squirrel* (blog), May 14, 2019. https://scientistseessquirrel.wordpress.com/2019/05/14/reclaiming-voice-in-scientific-writing/.

———. 2019b. "Reconciling the Two Functions of Peer Review." *Scientist Sees Squirrel* (blog), March 25, 2019. https://scientistseessquirrel.wordpress.com/2019/03/25/reconciling-the-two-functions-of-peer-review/.

———. 2020a. *Charles Darwin's Barnacle and David Bowie's Spider: How Scientific Names Celebrate Adventurers, Heroes, and Even a Few Scoundrels*. New Haven, CT: Yale University Press.

———. 2020b. "Moby Dick and Scientific Writing." *Scientist Sees Squirrel* (blog), August 11, 2020. https://scientistseessquirrel.wordpress.com/2020/08/11/moby-dick-and-scientific-writing/.

———. 2020c. "Steal This (Updated) Syllabus for Scientific Writing." *Scientist Sees Squirrel* (blog), May 26, 2020. https://scientistseessquirrel.wordpress.com/2020/05/26/steal-this-updated-syllabus-for-scientific-writing/.

———. 2021. "Blogging, Writing Practice, and Self-Discovery." *Scientist Sees Squirrel* (blog), June 22, 2021. https://scientistseessquirrel.wordpress.com/2021/06/22/blogging-writing-practice-and-self-discovery/.

———. 2022a. "Effective Grant Proposals, Part 5: Yes, Do Sweat the Small Stuff." *Scientist Sees Squirrel* (blog), July 19, 2022. https://scientistseessquirrel.wordpress.com/2022/07/19/effective-grant-proposals-part-4-yes-do-sweat-the-small-stuff/.

———. 2022b. "How Circular Expectations Damage Our Scientific Writing." *Scientist Sees Squirrel* (blog), September 27, 2022. https://scientistseessquirrel.wordpress.com/2022/09/27/how-circular-expectations-damage-our-scientific-writing/.

———. 2022c. *The Scientist's Guide to Writing: How to Write More Easily and Effectively throughout Your Scientific Career*. 2nd ed. Princeton, NJ: Princeton University Press.

———. 2023a. "ChatGPT: Author, Acknowledgement, Method, or Tool?" *Scientist Sees Squirrel* (blog), May 23, 2023. https://scientistseessquirrel.wordpress.com/2023/05/23/chatgpt-author-acknowledgement-method-or-tool/.

———. 2023b. "SOURs: Strong Opinions Unmoored from Rationale." *Scientist Sees Squirrel* (blog), March 14, 2023. https://scientistseessquirrel.wordpress.com/2023/03/14/sours-strong-opinions-unmoored-from-rationale/.

———. 2024a. "ChatGPT, Teaching, and Writing Style (and Also, Fun)." *Scientist Sees Squirrel* (blog), October 15, 2024. https://scientistseessquirrel.wordpress.com/2024/10/15/chatgpt-teaching-and-writing-style-and-also-fun/.

———. 2024b. "Scientific Papers and Mystery Novels Are Two Different Things—but Advice about Introductions Often Disagrees." *Scientist Sees Squirrel* (blog), February 27, 2024. https://scientistseessquirrel.wordpress.com/2024/02/27/scientific-papers-and-mystery-novels-are-two-different-things-but-advice-about-introductions-often-disagrees/.

———. 2024c. "Why 'Scholar' Is Such a Great Word." *Scientist Sees Squirrel* (blog), February 13, 2024. https://scientistseessquirrel.wordpress.com/2024/02/13/why-scholar-is-such-a-great-word/.

———. 2025. "Making ChatGPT Write Badly (for Use in Teaching): An Example, and Lessons." *Scientist Sees Squirrel* (blog), July 8. https://scientistseessquirrel.wordpress.com/2025/07/08/making-chatgpt-write-badly-for-use-in-teaching-an-example-and-lessons/.

Heard, Stephen B., Chloe C. Cull, and Easton R. White. 2023. "If This Title Is Funny, Will You Cite Me? Citation Impacts of Humour and Other Features of Article

Titles in Ecology and Evolution." *FACETS* 8 (1): 1–15. https://doi.org/doi.org/10.1139/facets-2022-0079.

Hessler, Michael, Daniel M. Pöpping, Hanna Hollstein, Hendrik Ohlenburg, Philip H. Arnemann, Christina Massoth, Laura M. Seidel, Alexander Zarbock, and Manuel Wenk. 2018. "Availability of Cookies during an Academic Course Session Affects Evaluation of Teaching." *Medical Education* 52 (10): 1064–72. https://doi.org/10.1111/medu.13627.

Higham, Nicholas J. 2020. *Handbook of Writing for the Mathematical Sciences*. 3rd ed. Philadelphia: Society for Industrial and Applied Mathematics.

Hinds, John. 1987. "Reader versus Writer Responsibility: A New Typology." In *Writing across Languages: Analysis of L2 Text*, edited by Ulla M. Connor and Robert B. Kaplan, 141–52. Reading, MA: Addison-Wesley.

Hofmann, Angie. 2022. *Scientific Writing and Communication*. 5th ed. New York: Oxford University Press.

Hofmann, Valentin, Pratyusha Ria Kalluri, Dan Jurafsky, and Sharese King. 2024. "AI Generates Covertly Racist Decisions about People Based On Their Dialect." *Nature* 633 (8028): 147–54. https://doi.org/10.1038/s41586-024-07856-5.

Holmes, Cassie. 2024. *Happier Hour: How to Spend Your Time for a Better, More Meaningful Life*. London: Penguin Life.

Huggingface.co. 2025. "AI Energy Score Leaderboard—a Hugging Face Space by AIEnergyScore." https://huggingface.co/spaces/AIEnergyScore/Leaderboard.

Hull, Akasha, Patricia Bell-Scott, and Barbara Smith, eds. 2015. *But Some of Us Are Brave: Black Women's Studies*. 2nd ed. New York: Feminist Press at CUNY.

Hull, Glynda, and Mike Rose. 1989. "Rethinking Remediation: Toward a Social-Cognitive Understanding of Problematic Reading and Writing." *Written Communication* 6 (2): 139–54. https://doi.org/10.1177/0741088389006002001.

Hyland, Fiona, and Ken Hyland. 2001. "Sugaring the Pill: Praise and Criticism in Written Feedback." *Journal of Second Language Writing* 10 (3): 185–212. https://doi.org/10.1016/S1060-3743(01)00038-8.

Ideland, Malin. 2018. "Science, Coloniality, and 'the Great Rationality Divide.'" *Science & Education* 27 (7): 783–803. https://doi.org/10.1007/s11191-018-0006-8.

Ingley, Spencer J., and Austin Pack. 2023. "Leveraging AI Tools to Develop the Writer Rather Than the Writing." *Trends in Ecology & Evolution* 38 (9): 785–87. https://doi.org/10.1016/j.tree.2023.05.007.

Irvine, Jeff. 2017. "A Comparison of Revised Bloom and Marzano's New Taxonomy of Learning." *Research in Higher Education Journal* 33 (November): 172608.

———. 2021. "Taxonomies in Education: Overview, Comparison, and Future Directions." *Journal of Education and Development* 5 (2): 1–25.

Iyengar, Shanto, and Douglas S. Massey. 2018. "Scientific Communication in a Post-truth Society." *Proceedings of the National Academy of Sciences* 116 (16): 7656–61. https://doi.org/10.1073/pnas.1805868115.

Jackson, Brian. 2015. "Watching Other People Teach: The Challenge of Classroom Observations." In *Assessing the Teaching of Writing: Twenty-First Century Trends and Technologies*, edited by Amy E. Dayton, 45–60. Logan: Utah State University Press.

Jacoby, Sally, and Elinor Ochs. 1995. "Co-construction: An Introduction." *Re-*

search on Language and Social Interaction 28 (3): 171–83. https://doi.org/10.1207/s15327973rlsi2803_1.

Jamieson, Sandra, and Rebecca Howard. 2019. "Rethinking the Relationship between Plagiarism and Academic Integrity." *Revue internationale des technologies en pédagogie universitaire / International Journal of Technologies in Higher Education* 16 (2): 69–85.

Januchowski-Hartley, Stephanie R., Natalie Sopinka, Bethann Garramon Merkle, Christina Lux, Anna Zivian, Patrick Goff, and Samantha Oester. 2018. "Poetry as a Creative Practice to Enhance Engagement and Learning in Conservation Science." *BioScience* 68 (11): 905–11. https://doi.org/10.1093/biosci/biy105.

Johns, Ann M. 2019. "Writing in the Interstices: Assisting Novice Undergraduates in Analyzing Authentic Writing Tasks." In *Changing Practices for the L2 Writing Classroom: Moving Beyond the Five-Paragraph Essay*, edited by Nigel A. Caplan and Ann M. Johns, 133–49. Ann Arbor: University of Michigan Press.

Johnson, Stephen. 2010. "Transferable Skills." In *The SAGE Handbook of Philosophy of Education*, edited by Robin Barrow, Richard Bailey, David Carr, and Christine McCarthy, 353–68. Thousand Oaks, CA: Sage Publications. https://sk.sagepub.com/reference/hdbk_philosophyeducation/n24.xml.

Joyce, James. 1939. *Finnegans Wake*. London: Faber and Faber.

Kahneman, Daniel. 2013. *Thinking, Fast and Slow*. New York: Farrar, Straus and Giroux.

Kamiya, George. 2020. "The Carbon Footprint of Streaming Video: Fact-Checking the Headlines." IEA, December 11, 2020. https://www.iea.org/commentaries/the-carbon-footprint-of-streaming-video-fact-checking-the-headlines.

Kearns, Faith. 2021. *Getting to the Heart of Science Communication: A Guide to Effective Engagement*. Washington, DC: Island Press.

Kelly, Erin E. 2023. "The Good, the Bad, and the Ugly of Peer Review." In *Writing Spaces*, edited by Trace Daniels-Lerberg, Dana L. Driscoll, Mary Stewart, and Matthew A. Vetter, 5:299–317. Anderson, SC: Parlor Press.

Kennell, Vicki, Amy Elliot, and Joshua Weirick. 2017. "Tinkering with Comments: Tailoring Practice by Spying on Written Artifacts." *Purdue Writing Lab / Purdue OWL Presentations*, no. 15 (November). https://docs.lib.purdue.edu/writinglabpres/15.

Kepner, Christine G. 1991. "An Experiment in the Relationship of Types of Written Feedback to the Development of Second-Language Writing Skills." *Modern Language Journal* 75 (3): 305–13. https://doi.org/10.2307/328724.

King, Stephen. 2020. *On Writing: A Memoir of the Craft*. Reissue ed. New York: Scribner.

Knight, Lucy. 2023. "Authors Call for AI Companies to Stop Using Their Work without Consent." *Guardian*, July 20, 2023. https://www.theguardian.com/books/2023/jul/20/authors-call-for-ai-companies-to-stop-using-their-work-without-consent.

Kobak, Dmitry, Rita González-Márquez, Emőke-Ágnes Horvát, and Jan Lause. 2025. "Delving into LLM-Assisted Writing in Biomedical Publications through Excess Vocabulary." *Science Advances* 11 (27): eadt3813. https://doi.org/10.1126/sciadv.adt3813.

Kotcher, John E., Teresa A. Myers, Emily K. Vraga, Neil Stenhouse, and Edward W.

Mailbach. 2017. "Does Engagement in Advocacy Hurt the Credibility of Scientists? Results from a Randomized National Survey Experiment." *Environmental Communication* 11 (3): 415–29. https://doi.org/10.1080/17524032.2016.1275736.

Krashen, Stephen D. 1982. *Principles and Practice in Second Language Acquisition*. New York: Pergamon Press.

Kubota, R. 1992. "Contrastive Rhetoric of Japanese and English: A Critical Approach." PhD thesis, University of Toronto, Ontario.

Lamott, Anne. 1994. *Bird by Bird: Some Instructions on Writing and Life*. New York: Pantheon.

Lang, James M. 2021. *Small Teaching: Everyday Lessons from the Science of Learning*. 2nd ed. Hoboken, NJ: Jossey-Bass.

Larsen, Tori M., Bianca H. Endo, Alexander T. Yee, Tony Do, and Stanley M. Lo. 2022. "Probing Internal Assumptions of the Revised Bloom's Taxonomy." *CBE—Life Sciences Education* 21 (4): ar66. https://doi.org/10.1187/cbe.20-08-0170.

Lee, David, and John Swales. 2006. "A Corpus-Based EAP Course for NNS Doctoral Students: Moving from Available Specialized Corpora to Self-Compiled Corpora." *English for Specific Purposes* 25 (1): 56–75. https://doi.org/10.1016/j.esp.2005.02.010.

Levis, John M., and Greta M. Levis. 2003. "A Project-Based Approach to Teaching Research Writing to Nonnative Writers." *IEEE Transactions on Professional Communication* 46 (3): 210–20. https://doi.org/10.1109/TPC.2003.816788.

Li, Mimi. 2021. *Researching and Teaching Second Language Writing in the Digital Age*. Cham, Switzerland: Palgrave Macmillan.

Li, Pengfei, Jianyi Yang, Mohammad A. Islam, and Shaolei Ren. 2023. "Making AI Less 'Thirsty': Uncovering and Addressing the Secret Water Footprint of Ai Models." arXiv. https://doi.org/10.48550/arXiv.2304.03271.

Li, Tingting, Emily Reigh, Peng He, and Emily Adah Miller. 2023. "Can We and Should We Use Artificial Intelligence for Formative Assessment in Science?" *Journal of Research in Science Teaching* 60 (6): 1385–89. https://doi.org/10.1002/tea.21867.

Liboiron, Max, Justine Ammendolia, Katharine Winsor, Alex Zahara, Hillary Bradshaw, Jessica Melvin, Charles Mather, et al. 2017. "Equity in Author Order: A Feminist Laboratory's Approach." *Catalyst: Feminism, Theory, Technoscience* 3 (2): 1–17. https://doi.org/10.28968/cftt.v3i2.28850.

Lillis, Theresa, and Mary Jane Curry. 2010. *Academic Writing in a Global Context: The Politics and Practices of Publishing in English*. Milton Park, UK: Routledge.

Lira, Benjamin, Todd Rogers, Daniel G. Goldstein, Lyle Ungar, and Angela L. Duckworth. 2025. "Learning Not Cheating: AI Assistance Can Enhance Rather than Hinder Skill Development." Preprint, arXiv, February 22. https://doi.org/10.48550/arXiv.2502.02880.

Lisman, Shiela A. 1979. "The Best of All Possible Worlds: Where X Replaces AWK." In *Classroom Practices in Teaching English 1979–1980: How to Handle the Paper Load*, edited by Gene Stanford, 103–5. Urbana, IL: National Council of Teachers of English.

Lorés-Sanz, Rosa. 2018. "Hybrid Rhetorical Structure in English Sociology Research

Article Abstracts: The Ambit of ELF and Translation." In *Intercultural Perspectives on Research Writing*, edited by Pilar Mur-Dueñas and Jolanta Šinkūnienė, 175–94. AILA Applied Linguistics Series. Amsterdam: John Benjamins.

Losh, Elizabeth, Jonathan Alexander, Kevin Cannon, and Zander Cannon. 2013. *Understanding Rhetoric: A Graphic Guide to Writing*. Boston: Bedford Books.

Macfarlane, Bruce. 2007. "Defining and Rewarding Academic Citizenship: The Implications for University Promotions Policy." *Journal of Higher Education Policy and Management* 29 (3): 261–73. https://doi.org/10.1080/13600800701457863.

MacKenzie, Ian. 2015. "Rethinking Reader and Writer Responsibility in Academic English." *Applied Linguistics Review* 6 (1): 1–21. https://doi.org/10.1515/applirev-2015-0001.

Marcus, Gary. 2025. "Why DO Large Language Models Hallucinate?" Substack newsletter. *Marcus on AI*, May 5. https://garymarcus.substack.com/p/why-do-large-language-models-hallucinate.

Marks, Laura U., and Radek Przedpełski. 2022. "The Carbon Footprint of Streaming Media: Problems, Calculations, Solutions." In *Film and Television Production in the Age of Climate Crisis: Towards a Greener Screen*, edited by Pietari Kääpä and Hunter Vaughan, 207–34. Cham: Springer International. https://doi.org/10.1007/978-3-030-98120-4_10.

Martineau, Kim. 2023. "What Is Retrieval-Augmented Generation (RAG)?" IBM Research, August 22, 2023. https://research.ibm.com/blog/retrieval-augmented-generation-RAG.

Martínez, Alejandro, and Stefano Mammola. 2021. "Specialized Terminology Reduces the Number of Citations of Scientific Papers." *Proceedings of the Royal Society B: Biological Sciences* 288 (1948): 20202581. https://doi.org/10.1098/rspb.2020.2581.

Masley, Andy. 2025a. "Using ChatGPT Is Not Bad for the Environment." Substack newsletter. *The Weird Turn Pro*, January 13. https://andymasley.substack.com/p/individual-ai-use-is-not-bad-for.

Masley, Andy. 2025b. "A Cheat Sheet for Why Using ChatGPT Is Not Bad for the Environment." Substack newsletter. *The Weird Turn Pro*, April 28. https://andymasley.substack.com/p/a-cheat-sheet-for-conversations-about.

Masters, Ken. 2023. "Medical Teacher's First ChatGPT's Referencing Hallucinations: Lessons for Editors, Reviewers, and Teachers." *Medical Teacher* 45 (7): 673–75. https://doi.org/10.1080/0142159X.2023.2208731.

Maxwell, Amanda, Guy J. Curtis, and Lucia Vardanega. 2008. "Does Culture Influence Understanding and Perceived Seriousness of Plagiarism?" *International Journal for Educational Integrity* 4 (2): 25–40. https://doi.org/10.21913/IJEI.v4i2.412.

Mazak, Cathy. 2019. "Combating Writing Guilt." *Academic Writing Amplified*, November 19. Episode 12. Available at https://open.spotify.com/episode/2ez0zQCSrXdXjpRc261SjV?si=Rcp7UqM5QqmHDYg7Xe6yNg.

Mbutho, Nozuko P., and Catherine Hutchings. 2021. "The Complex Concept of Plagiarism: Undergraduate and Postgraduate Student Perspectives." *Perspectives in Education* 39 (2): 67–81.

McGlynn, Terry. 2020. *The Chicago Guide to College Science Teaching*. 1st ed. Chicago: University of Chicago Press.

McNeilly, Anne. 2014. "Minimal Marking: A Success Story." *Canadian Journal for the Scholarship of Teaching and Learning* 5 (1). https://doi.org/10.5206/cjsotl-rcacea.2014.1.7.

Melville, Herman. 1851. *Moby-Dick; or, The Whale*. New York: Harper and Brothers.

Mengel, Friederike, Jan Sauermann, and Ulf Zölitz. 2019. "Gender Bias in Teaching Evaluations." *Journal of the European Economic Association* 17 (2): 535–66. https://doi.org/10.1093/jeea/jvx057.

Merkle, Bethann Garramon. 2019. "The Chicago Guide to Communicating Science, 2nd Ed. (Book Review)." *Journal of Wildlife Management* 83 (3): 742–43. https://doi.org/10.1002/jwmg.21606.

———. 2022. "Writing Science: Leveraging the Annotated Bibliography as a Writing Tool." *Bulletin of the Ecological Society of America* 103 (1): e01936. https://doi.org/10.1002/bes2.1936.

———. 2023a. "Can ChatGPT Write a #SciComm Grant Proposal?" *School of Good Trouble* (blog), March 3, 2023. https://www.commnatural.com/post/can-chatgpt-write-a-scicomm-grant-proposal.

———. 2023b. "Facilitated Road Maps: Coaching Students to Embed Planning in Major Communication Projects." In *Teaching Science Students to Communicate: A Practical Guide*, edited by Susan Rowland and Louise Kuchel, 495–504. Cham, Switzerland: Springer.

———. 2023c. "Writing like Broccoli: Disguise the Effort so Students Can Get Past Hang-Ups + Be Better Writers." *School of Good Trouble* (blog), July 25, 2023. https://www.commnatural.com/post/like-broccoli-better-student-writing.

———. 2024a. "Frame Your Own Intentions for Your Teaching and Then Document How You Are Achieving Them (or, If Student Evals Suck, How Do We Know We're Doing a Good Job Teaching?)." *School of Good Trouble* (blog), January 16, 2024. https://www.commnatural.com/post/you-decide-what-is-good-teaching.

———. 2024b. "Making "TEA" Can Improve Students' Writing." *School of Good Trouble* (blog), March 26, 2024. https://www.commnatural.com/post/tea-improves-student-writing.

———. 2024c. "Soulless Academic Writing Happens on Purpose." *School of Good Trouble* (blog), March 5, 2024, updated April 6, 2024. https://www.commnatural.com/post/soulless-academic-writing-happens-on-purpose.

———. 2024d. "Three Things You Can Tell Someone When Their Idea Is Good and the Writing Is Competent, but the Draft Is Not Working." *School of Good Trouble* (blog), February 6, 2024. https://www.commnatural.com/post/writings-competent-but-not-working-now-what.

———. 2025. "A Scicomm Take on Why Debates Feel Like a Dead End." *School of Good Trouble* (blog), January 29, 2025. https://www.commnatural.com/post/debates-are-a-dead-end.

Merkle, Bethann Garramon, Evelyn Valdez-Ward, Priya Shukla, and Skylar Bayer. 2022. "Sharing Science through Shared Values, Goals, and Stories: An Evidence-Based Approach to Making Science Matter." *Human-Wildlife Interactions* 15 (3): 598–614. https://doi.org/10.26077/9wss-av78.

Mewburn, Inger. 2013. "How to Create 'Authoritative Voice' in Your Writing." *Thesis Whisperer* (blog), July 2, 2013. https://thesiswhisperer.com/2013/07/03/how-to-create-authoritative-voice-in-your-writing/.

——. 2020. "Why Academic Writing Sucks (and How We Can Fix It)." *Thesis Whisperer* (blog), June 9, 2020. https://thesiswhisperer.com/2020/06/10/why-academic-writing-sucks-and-how-we-can-fix-it/.

——. 2021. "How to Finish That Big Writing Project (and Get On with Your Life)." *Thesis Whisperer* (blog), October 5, 2021. https://thesiswhisperer.com/2021/10/06/a-few-organisation-tips/.

——. 2024. "Why You Need a Nobot." *Thesis Whisperer* (blog), October 31, 2024. https://thesiswhisperer.com/2024/10/31/why-you-need-a-nobot/.

——. 2025. "Getting Good Feedback during the Academic Apocalypse." *Thesis Whisperer* (blog), February 5, 2025. https://thesiswhisperer.com/2025/02/05/getting-good-feedback-during-the-academic-apocalypse/.

Michel, Jean-Baptiste, Yuan Kui Shen, Aviva Presser Aiden, Adrian Veres, Matthew K. Gray, The Google Books Team, Joseph P. Pickett, et al. 2011. "Quantitative Analysis of Culture Using Millions of Digitized Books." *Science* 331 (6014): 176–82. https://doi.org/10.1126/science.1199644.

Misra, Joy, Jennifer H. Lundquist, Elissa Holmes, and Stephanie Agiomavritis. 2011. "The Ivory Ceiling of Service Work." AAUP, January 4, 2011. https://www.aaup.org/article/ivory-ceiling-service-work.

Mlynarek, Julia J., Chandra E. Moffat, Sara Edwards, Anthony L. Einfeldt, Allyson Heustis, Rob Johns, Mallory MacDonnell, et al. 2017. "Enemy Escape: A General Phenomenon in a Fragmented Literature?" *FACETS* 2 (January): 1015–44.

Montgomery, Beronda L. 2019. "How I Work and Thrive in Academia—from Affirmation, Not for Affirmation." *Being Lazy and Slowing Down* (blog), September 30, 2019. https://lazyslowdown.com/how-i-work-and-thrive-in-academia-from-affirmation-not-for-affirmation/.

——. 2021. *Lessons from Plants*. Cambridge, MA: Harvard University Press.

Moore, Cindy. 2015. "Administrative Priorities and the Case for Multiple Methods." In *Assessing the Teaching of Writing: Twenty-First Century Trends and Technologies*, edited by Amy E. Dayton, 133–51. Logan: Utah State University Press.

Mott, Carrie, and Daniel Cockayne. 2017. "Citation Matters: Mobilizing the Politics of Citation toward a Practice of 'Conscientious Engagement.'" *Gender, Place & Culture* 24 (7): 954–73. https://doi.org/10.1080/0966369X.2017.1339022.

Mowatt, Rasul A. 2019. "Twelve Years a Servant: Race and the Student Evaluation of Teaching." *SCHOLE: A Journal of Leisure Studies and Recreation Education* 34 (2): 109–19. https://doi.org/10.1080/1937156X.2019.1622949.

Muir, Tom, and Kristin Solli. 2019. "The Unreal and the Real: English for Research Purposes in Norway." In *Pedagogies and Policies for Publishing Research in English*, edited by David Ian Hanauer, Karen Englander, and Laura-Mihaela Muresan, 91–106. Milton Park, UK: Routledge.

Mur-Dueñas, Pilar, and Jolanta Šinkūnienė, eds. 2018. *Intercultural Perspectives on Research Writing*. AILA Applied Linguistics Series. Amsterdam: John Benjamins.

Murphy, Julie A., and Anne Shelley. 2020. "Textbook Affordability in the Time of

Covid-19." *Serials Review* 46 (3): 232–37. https://doi.org/10.1080/00987913.2020.1806656.

Murray, Dakota, Clara Boothby, Huimeng Zhao, Vanessa Minik, Nicolas Bérubé, Vincent Larivière, and Cassidy R. Sugimoto. 2020. "Exploring the Personal and Professional Factors Associated with Student Evaluations of Tenure-Track Faculty." *PLOS ONE* 15 (6): e0233515. https://doi.org/10.1371/journal.pone.0233515.

Murray, Darrin S. 2019. "The Precarious New Faculty Majority: Communication and Instruction Research and Contingent Labor in Higher Education." *Communication Education* 68 (2): 235–45. https://doi.org/10.1080/03634523.2019.1568512.

Nabokov, Vladimir. 1955. *Lolita*. Paris: Olympia Press.

Nägele, Christof, and Barbara E. Stalder. 2017. "Competence and the Need for Transferable Skills." In *Competence-Based Vocational and Professional Education: Bridging the Worlds of Work and Education*, edited by Martin Mulder, 739–53. New York: Springer.

Narayanan, Arvind, and Sayash Kapoor. 2024. *AI Snake Oil: What Artificial Intelligence Can Do, What It Can't, and How to Tell the Difference*. Princeton: Princeton University Press.

Nature. 2023. "Tools Such as ChatGPT Threaten Transparent Science; Here Are Our Ground Rules for Their Use." Editorial, *Nature* 613 (7945): 612–12. https://doi.org/10.1038/d41586-023-00191-1.

Nauman, Sarwat. 2019. "The Impact of English Language Teaching Reforms on Pakistani Scholars' Language and Research Skills." In *Pedagogies and Policies for Publishing Research in English*, edited by David Ian Hanauer, Karen Englander, and Laura-Mihaela Muresan, 179–92. Milton Park, UK: Routledge.

Nazmi, Aydin, Suzanna Martinez, Ajani Byrd, Derrick Robinson, Stephanie Bianco, Jennifer Maguire, Rashida M. Crutchfield, Kelly Condron, and Lorrene Ritchie. 2019. "A Systematic Review of Food Insecurity among US Students in Higher Education." *Journal of Hunger & Environmental Nutrition* 14 (5): 725–40. https://doi.org/10.1080/19320248.2018.1484316.

NCES National Center for Education Statistics. 2023. "Characteristics of Postsecondary Faculty. *Condition of Education*." US Department of Education, Institute of Education Sciences.

Newkirk, Thomas. 2014. *Minds Made for Stories: How We Really Read and Write Informational and Persuasive Texts*. Portsmouth, NH: Heinemann.

Newport, Cal. 2016. *Deep Work: Rules for Focused Success in a Distracted World*. New York: Grand Central Publishing.

Nijhuis, Michelle. 2016. *The Science Writers' Essay Handbook: How to Craft Compelling True Stories in Any Medium*. Published by the author.

Nikolaus, Cassandra J., Ruopeng An, Brenna Ellison, and Sharon M. Nickols-Richardson. 2020. "Food Insecurity among College Students in the United States: A Scoping Review." *Advances in Nutrition (Bethesda, Md.)* 11 (2): 327–48. https://doi.org/10.1093/advances/nmz111.

Noy, Shakked, and Whitney Zhang. 2023. "Experimental Evidence on the Productivity Effects of Generative Artificial Intelligence." *Science* 381 (6654): 187–92. https://doi.org/10.1126/science.adh2586.

O'Donnell, James. 2025. "DeepSeek Might Not Be Such Good News for Energy after All." *MIT Technology Review*, January 31, 2025. https://www.technologyreview.com/2025/01/31/1110776/deepseek-might-not-be-such-good-news-for-energy-after-all/.

Olesen, Kristoffer B., Mette K. Christensen, and Lotte D. O'Neill. 2021. "What Do We Mean by 'Transferable Skills'? A Literature Review of How the Concept Is Conceptualized in Undergraduate Health Sciences Education." *Higher Education, Skills and Work-Based Learning* 11 (3): 616–34.

Oliviera, Osvaldo N., Jr., Ethel Schuster, Sandra M. Aluísio, and Haim Levkowitz. 2014. "Reading, Annotating, Compiling, and Producing Text for Scientific Papers." In *Writing Scientific Papers in English Successfully: Your Complete Roadmap*, 57–82. Andover, MA: Hypertek.com.

Omobowale, Ayokunle Olumuyiwa, Olayinka Akanle, and Charles Akinsete. 2019. "Scholarly Publishing in Nigeria: The Enduring Effects of Colonization." In *Pedagogies and Policies for Publishing Research in English*, edited by David Ian Hanauer, Karen Englander, and Laura-Mihaela Muresan, 215–32. Milton Park, UK: Routledge.

O'Neill, Kathryn K., Kerith J. Conron, Abbie E. Goldberg, and Rubeen Guardado. 2022. "Experiences of LGBTQ People in Four-Year Colleges and Graduate Programs: Findings from a National Probability Survey." UCLA School of Law—Williams Institute. https://williamsinstitute.law.ucla.edu/publications/lgbtq-colleges-grad-school/.

OpenAI. 2025. "OpenAI O3 and O4-Mini System Card." https://openai.com/index/o3-o4-mini-system-card/.

Oruç, A. Yavuz. 2011. *Handbook of Scientific Proposal Writing*. Boca Raton, FL: CRC Press.

Pajares, Frank. 2003. "Self-Efficacy Beliefs, Motivation, and Achievement in Writing: A Review of the Literature." *Reading & Writing Quarterly* 19 (2): 139–58. https://doi.org/10.1080/10573560308222.

Panadero, Ernesto, and Anders Jonsson. 2020. "A Critical Review of the Arguments against the Use of Rubrics." *Educational Research Review* 30 (June): 100329. https://doi.org/10.1016/j.edurev.2020.100329.

Parnther, Ceceilia. 2020. "Academic Misconduct in Higher Education: A Comprehensive Review." *Journal of Higher Education Policy and Leadership Studies* 1 (1): 25–45. https://doi.org/10.29252/johepal.1.1.25.

Pennycook, Alastair. 1996. "Borrowing Others' Words: Text, Ownership, Memory, and Plagiarism." *TESOL Quarterly* 30 (2): 201–30. https://doi.org/10.2307/3588141.

Penrose, Ann M., and Steven B. Katz. 2009. *Writing in the Sciences: Exploring Conventions of Scientific Discourse*. 3rd ed. New York: Pearson.

Pérez-Llantada, Carmen, Ramón Plo, and Gibson R. Ferguson. 2011. "'You Don't Say What You Know, Only What You Can': The Perceptions and Practices of Senior Spanish Academics Regarding Research Dissemination in English." *English for Specific Purposes* 30 (1): 18–30. https://doi.org/10.1016/j.esp.2010.05.001.

Perrigo, Billy. 2023. "OpenAI Used Kenyan Workers on Less Than $2 Per Hour." *Time*, January 18, 2023. https://time.com/6247678/openai-chatgpt-kenya-workers/.

Petrat-Melin, Bjørn, and Svend Dam. 2023. "Textural and Consumer-Aided Characterisation and Acceptability of a Hybrid Meat and Plant-Based Burger Patty." *Foods* 12 (11): 2246. https://doi.org/10.3390/foods12112246.

Pickering, John W, and Garry Hornby. 2005. "Plagiarism and International Students: A Matter of Values Differences?" In *16th ISANA International Education Conference*, November 29–December 2, 2005, Christchurch, New Zealand.

Pineda, Ana. 2020. "The Two Things You Can Do to Get Better at Writing." *I Focus and Write* (blog), April 7, 2020. https://www.ifocusandwrite.com/post/readingandwriting.

Pitkäniemi, Harri. 2010. "How the Teacher's Practical Theory Moves to Teaching Practice: A Literature Review and Conclusions." *Education Inquiry* 1 (3): 157–75.

Plavén-Sigray, Pontus, Granville J. Matheson, Björn C. Schiffler, and William H. Thompson. 2017. "The Readability of Scientific Texts Is Decreasing over Time." *eLife* 6 (September): e27725. https://doi.org/10.7554/eLife.27725.

Polfus, Jean, Deborah Simmons, Michael Neyelle, Walter Bayha, Frederick Andrew, Leon Andrew, Bethann Garramon Merkle, Keren Rice, and Micheline Manseau. 2017. "Creative Convergence: Exploring Biocultural Diversity through Art." *Ecology and Society* 22 (2): 4. https://doi.org/10.5751/ES-08711-220204.

Prescod-Weinstein, Chanda. 2021. *The Disordered Cosmos: A Journey into Dark Matter, Spacetime, and Dreams Deferred*. Rep. ed. New York: Bold Type Books.

Purcell, Kristen, Judy Buchanan, and Linda Friedrich. 2013. *The Impact of Digital Tools on Student Writing and How Writing Is Taught in Schools*. Pew Research Center. https://www.pewresearch.org/internet/2013/07/16/the-impact-of-digital-tools-on-student-writing-and-how-writing-is-taught-in-schools/.

Qi, Xiukun, and Lida Liu. 2007. "Differences between Reader/Writer Responsible Languages Reflected in EFL Learners' Writing." *Intercultural Communication Studies* 16 (3): 148–59.

Radford, Alexandria W., Melissa Cominole, and Paul Skomsvold. 2015. "Demographic and Enrollment Characteristics of Nontraditional Undergraduates: 2011–12." NCES 2015025. National Center for Education Statistics. https://nces.ed.gov/pubsearch/pubsinfo.asp?pubid=2015025.

Rafoth, Ben. 2010. "Why Visit Your Campus Writing Center?" In *Writing Spaces: Readings on Writing*, edited by Charles Lowe and Pavel Zemliansky, 1:146–55. Anderson, SC: Parlor Press.

Ragupathi, Kiruthika, and Adrian Lee. 2020. "Beyond Fairness and Consistency in Grading: The Role of Rubrics in Higher Education." In *Diversity and Inclusion in Global Higher Education: Lessons from across Asia*, edited by Catherine Shea Sanger and Nancy W. Gleason, 73–95. Singapore: Springer.

Rankin, Tess. 2022. "How to Develop Your Authorial Voice in Academic Writing." *Flatpage* (blog), October 24, 2022. https://flatpage.com/how-to-develop-your-authorial-voice-in-academic-writing/.

Reed, Mark S., Bethann Garramon Merkle, Elizabeth J. Cook, Caitlin Hafferty, Adam P. Hejnowicz, Richard Holliman, Ian D. Marder, et al. 2024. "Reimagining the Language of Engagement in a Post-stakeholder World." *Sustainability Science* 19 (4): 1481–90.Reynders, Gil, Juliette Lantz, Suzanne M. Ruder, Courtney L. Stanford, and Renée S. Cole. 2020. "Rubrics to Assess Critical Thinking and Information Processing in Undergraduate STEM Courses." *International Journal of STEM Education* 7 (1): 9. https://doi.org/10.1186/s40594-020-00208-5.

Rillig, Matthias C., Marlene Ågerstrand, Mohan Bi, Kenneth A. Gould, and Uli Sauerland. 2023. "Risks and Benefits of Large Language Models for the Environment." *Environmental Science & Technology* 57 (9): 3464–66. https://doi.org/10.1021/acs.est.3c01106.

Ritchie, Hannah. 2024. "What's the Carbon Footprint of Using ChatGPT?" *Sustainability by Numbers*, November 18. https://www.sustainabilitybynumbers.com/p/carbon-footprint-chatgpt.

Robbins, Susan P. 2016. "Finding Your Voice as an Academic Writer (and Writing Clearly)." *Journal of Social Work Education* 52 (2): 133–35. https://doi.org/10.1080/10437797.2016.1151267.

Roberts, David Lindsay. 2014. "History of Tools and Technologies in Mathematics Education." In *Handbook on the History of Mathematics Education*, edited by Alexander Karp and Gert Schubring, 565–78. New York: Springer.

Rockquemore, Kerry Ann. 2010. "Shut Up and Write." June 13, 2010. https://www.insidehighered.com/advice/2010/06/14/shut-and-write.

Rodgers, Carol. 2002. "Defining Reflection: Another Look at John Dewey and Reflective Thinking." *Teachers College Record: The Voice of Scholarship in Education* 104 (4): 842–66. https://doi.org/10.1111/1467-9620.0018.

Rose, Mike. 1985. "The Language of Exclusion: Writing Instruction at the University." *College English* 47 (4): 341–59. https://doi.org/10.2307/376957.

Ross, Elizabeth. 2018. "Learning Disability or Learning Difference?" *SMARTS* (blog), June 13, 2018. https://smarts-ef.org/blog/learning-disabilities-learning-differences/.

Rowland, Susan, and Louise Kuchel, eds. 2023. *Teaching Science Students to Communicate: A Practical Guide*. Cham, Switzerland: Springer.

Russell, David R. 2002. *Writing in the Academic Disciplines: A Curricular History*. 2nd ed. Carbondale: Southern Illinois University Press.

Ryan, Mary, and Michael Ryan. 2015. "A Model for Reflection in the Pedagogic Field of Higher Education." In *Teaching Reflective Learning in Higher Education: A Systematic Approach Using Pedagogic Patterns*, 15–27. New York: Springer.

Sagan, Carl. 1985. *Contact*. New York: Simon and Schuster.

Salemi, Alireza, and Hamed Zamani. 2024. "Evaluating Retrieval Quality in Retrieval-Augmented Generation." In *Proceedings of the 47th International ACM SIGIR Conference on Research and Development in Information Retrieval*, 2395–2400. SIGIR '24. New York: Association for Computing Machinery. https://doi.org/10.1145/3626772.3657957.

Salvaggio, Eryk. 2025. "In Copyright 'Wins' for Anthropic and Meta, Judges Leave Ample Room for Future Defeats." Tech Policy Press, June 26. https://techpolicy.press/in-copyright-wins-for-anthropic-and-meta-judges-leave-ample-room-for-future-defeats.

Sanders-Reio, Joanne, Patricia A. Alexander, Thomas G. Reio, and Isadore Newman. 2014. "Do Students' Beliefs about Writing Relate to Their Writing Self-Efficacy, Apprehension, and Performance?" *Learning and Instruction* 33 (October): 1–11. https://doi.org/10.1016/j.learninstruc.2014.02.001.

Sarewitz, Daniel. 2004. "How Science Makes Environmental Controversies Worse." *Environmental Science & Policy* 7 (5): 385–403. https://doi.org/10.1016/j.envsci.2004.06.001.

Saville-Troike, Muriel, and Karen Barto. 2016. *Introducing Second Language Acquisition*. 3rd ed. Cambridge, UK: Cambridge University Press.

Scheufele, Dietram. 2013. "Communicating Science in Social Settings." *Proceedings of the National Academy of Sciences* 110: 14040–47. https://doi.org/10.1073/pnas.1213275110.

Schien, Daniel, Paul Shabajee, Huseyin Burak Akyol, Luke Benson, and Angeliki Katsenou. 2024. "Assessing the Carbon Reduction Potential for Video Streaming from Short-Term Coding Changes." In *2024 16th International Conference on Quality of Multimedia Experience (QoMEX)*, 22–28. https://doi.org/10.1109/QoMEX61742.2024.10598286.

Schimel, Joshua. 2011. *Writing Science: How to Write Papers That Get Cited and Proposals That Get Funded*. Oxford, UK: Oxford University Press.

Schuster, Ethel, Haim Levkowitz, and Osvaldo N. Jr. Oliveira, eds. 2014. *Writing Scientific Papers in English Successfully: Your Complete Roadmap*. Andover, MA: hyprtek.com, inc.

Schwalm, David E. 1985. "Degree of Difficulty in Basic Writing Courses: Insights from the Oral Proficiency Interview Testing Program." *College English* 47 (6): 629–40. https://doi.org/10.2307/377165.

Scrimgeour, Garry J., and Shelley D. Pruss. 2016. "Writing Highly Effective Reviews of a Scientific Manuscript." *Freshwater Science* 35 (4): 1076–81. https://doi.org/10.1086/688856.

Settles, Isis H., Martinque K. Jones, NiCole T. Buchanan, and Sheila T. Brassel. 2022. "Epistemic Exclusion of Women Faculty and Faculty of Color: Understanding Scholar(ly) Devaluation as a Predictor of Turnover Intentions." *Journal of Higher Education* 93 (1): 31–55. https://doi.org/10.1080/00221546.2021.1914494.

Shahriari, Hesamoddin, and Behzad Ghonsooly. 2019. "Examining the Status Quo of Publication in Iranian Higher Education: Perceptions and Strategies." In *Pedagogies and Policies for Publishing Research in English*, edited by David Ian Hanauer, Karen Englander, and Laura-Mihaela Muresan, 235–51. Milton Park, UK: Routledge.

Shaughnessy, Mina P. 1977. *Errors and Expectations: A Guide for the Teacher of Basic Writing*. New York: Oxford University Press.

Sheffield, Jenna P. 2016. "Thinking beyond Tools: Writing Program Administration and Digital Literacy." *Computers and Composition Online*, Fall 15–Fall 16. http://cconlinejournal.org/sheffield/index.html.

Silver, Naomi, Matthew Kaplan, Danielle LaVaque-Manty, and Deborah Meizlish, eds. 2013. *Using Reflection and Metacognition to Improve Student Learning: Across the Disciplines, across the Academy*. New York: Routledge.

Silvia, Paul J. 2014. *Write It Up: Practical Strategies for Writing and Publishing Journal Articles*. Washington, DC: American Psychological Association.

——. 2018. *How to Write a Lot: A Practical Guide to Productive Academic Writing*. Washington, DC: APA LifeTools.

Simis, Molly J., Haley Madden, and Sara K. Yeo. 2016. "The Lure of Rationality: Why

Does the Deficit Model Persist in Science Communication?" *Public Understanding of Science* 25 (4): 400–414. https://doi.org/10.1177/0963662516629749.

Singh, Gerald G. 2022. "Prestige Risks Homogenizing and Hampering Academia." *Nature* 610 (7933): 630–630. https://doi.org/10.1038/d41586-022-03406-z.

Smith, Anna Nellis B., and Bethann Garramon Merkle. 2021. "Meaning-Making in Science Communication: A Case for Precision in Word Choice." *Bulletin of the Ecological Society of America* 102 (1): e01794. https://doi.org/10.1002/bes2.1794.

Smith, Gabriel R., Carolina Bello, Lalasia Bialic-Murphy, Emily Clark, Camille S. Delavaux, Camille F. de Lauriere, Johan van den Hoogan, et al. 2024. "Ten Simple Rules for Using Large Language Models in Science, Version 1.0." *PLOS Computational Biology* 20 (1): e1011767. https://doi.org/10.1371/journal.pcbi.1011767.

Smith, Theo, and Amanda Kirby. 2021. *Neurodiversity at Work: Drive Innovation, Performance and Productivity with a Neurodiverse Workforce*. London: Kogan Page.

Snow, C. P. 1959. *The Two Cultures and the Scientific Revolution*. New York: Cambridge University Press.

Stanny, Claudia J. 2016. "Reevaluating Bloom's Taxonomy: What Measurable Verbs Can and Cannot Say about Student Learning." *Education Sciences* 6 (4): 37.

Steiss, Jacob, Tamara Tate, Steve Graham, Jazmin Cruz, Michael Hebert, Jiali Wang, Youngsun Moon, Waverly Tseng, Mark Warschauer, and Carol Booth Olson. 2024. "Comparing the Quality of Human and ChatGPT Feedback of Students' Writing." *Learning and Instruction* 91 (June):101894. https://doi.org/10.1016/j.learninstruc.2024.101894.

Stöffelbauer, Andreas. 2023. "How Large Language Models Work." *Data Science at Microsoft*, October 24. https://medium.com/data-science-at-microsoft/how-large-language-models-work-91c362f5b78f.

Swales, John M. 1990. *Genre Analysis: English in Academic and Research Settings*. 1st ed. Cambridge, UK: Cambridge University Press.

———. 2004. *Research Genres: Explorations and Applications*. Cambridge, UK: Cambridge University Press.

Swales, John M., and Christine Feak. 2012. *Academic Writing for Graduate Students: Essential Tasks and Skills*. 3rd ed. Ann Arbor: University of Michigan Press.

Sword, Helen. 2012. *Stylish Academic Writing*. Illus. ed. Cambridge, MA: Harvard University Press.

———. 2016a. "'Write Every Day!' A Mantra Dismantled." *International Journal for Academic Development* 21 (4): 312–22. https://doi.org/10.1080/1360144X.2016.1210153.

———. 2016b. *The Writer's Diet: A Guide to Fit Prose*. 2nd ed. Chicago: University of Chicago Press.

———. 2017. *Air & Light & Time & Space: How Successful Academics Write*. Illus. ed. Cambridge, MA: Harvard University Press.

———. 2023. *Writing with Pleasure*. Princeton, NJ: Princeton University Press.

Tagnin, Stella E.O. 2014. "Using Corpus Linguistics to Overcome the Language Barrier." In *Writing Scientific Papers in English Successfully*, edited by Ethel Schuster, Haim Levkowitz, and Osvaldo N. Jr. Oliveira, 83–114. Andover, MA: Hypertek.com.

Talbert, Robert. 2017. *Flipped Learning*. Sterling, VA: Routledge.

———. 2023. "Ungrading in STEM Courses." *Zeal: A Journal for the Liberal Arts* 1 (2): 112–16.

Thomson, Pat. 2017. "Co-writing with Your Supervisor—the Authorship Question." *Patter* (blog), April 10, 2017. https://patthomson.net/2017/04/10/co-writing-with-your-supervisor-the-authorship-question/.

Tolman, Anton O., and Janine Kremling, eds. 2016. *Why Students Resist Learning: A Practical Model for Understanding and Helping Students*. 1st ed. Sterling, VA: Routledge.

Tomlinson, Bill, Rebecca W. Black, Donald J. Patterson, and Andrew W. Torrance. 2024. "The Carbon Emissions of Writing and Illustrating Are Lower for AI than for Humans." *Scientific Reports* 14 (1): 3732.

Townsend, Martha A. 2016. "What Are Writing across the Curriculum and Writing in the Disciplines?" In *A Rhetoric for Writing Program Administrators*, 2nd ed., edited by Rita Malenczyk, 77–90. Anderson, SC: Parlor Press.

Trachtenberg, Jordan, Hwa Young Lee, Cheryl Anderson, Shine Chang, and Carrie Cameron. 2018. "Do You Speak Science? Dialect and Its Role in Research Training." *Understanding Interventions* 9 (1). https://www.understandinginterventionsjournal.org/article/3729.

Treglia, Maria O. 2008. "Feedback on Feedback: Exploring Student Responses to Teachers' Written Commentary." *Journal of Basic Writing* 27 (1): 105–37.

Tremain, Lisa. 2019. "(Dis)Positioning Writing Confidence, Reflecting on Writer Identity: A Writing about Writing Curriculum Aimed at Knowledge Transfer." In *Next Steps: New Directions for/in Writing about Writing*, edited by Barbara Bird, Doug Downs, I. Moriah McCracken, and Jan Rieman, 56–67. Logan: Utah State University Press.

———. 2023. "What Can I Add to the Discourse Communities? How Writers Use Code-Meshing and Translanguaging to Negotiate Discourse." In *Writing Spaces: Readings on Writing*, edited by Dana Driscoll, Mary Stewart, and Matthew Vetter, 5:87–101. Anderson, SC: Parlor Press.

Treves, Adrian. 2019. "Scientific Ethics and the Illusion of Naïve Objectivity." *Frontiers in Ecology and the Environment* 17 (7): 363. https://doi.org/10.1002/fee.2091.

Truscott, John, and Angela Y.-P. Hsu. 2008. "Error Correction, Revision, and Learning." *Journal of Second Language Writing* 17 (4): 292–305. https://doi.org/10.1016/j.jslw.2008.05.003.

Truss, Lynne. 2006. *Eats, Shoots and Leaves*. New York: Gotham.

Tseng, Waverly, and Mark Warschauer. 2023. "AI-Writing Tools in Education: If You Can't Beat Them, Join Them." *Journal of China Computer-Assisted Language Learning* 3 (2): 258–62. https://doi.org/10.1515/jccall-2023-0008.

Turbek, Sheela P., Taylor M. Chock, Kyle Donahue, Caroline A. Havrilla, Angela M. Oliverio, Stephanie K. Polutchko, Lauren G. Shoemaker, and Lara Vimercati. 2016. "Scientific Writing Made Easy: A Step-by-Step Guide to Undergraduate Writing in the Biological Sciences." *Bulletin of the Ecological Society of America* 97 (4): 417–26.

Turner, Ben. 2025. "Why Is DeepSeek Such a Game-Changer? Scientists Explain How the AI Models Work and Why They Were So Cheap to Build." *Livescience.Com*,

January 31, 2025. https://www.livescience.com/technology/artificial-intelligence/why-is-deekspeek-such-a-game-changer-scientists-explain-how-the-ai-models-work-and-why-they-were-so-cheap-to-build.

United States Copyright Office. 2025a. *Copyright and Artificial Intelligence: Part 2: Copyrightability*. Washington, DC: United States Copyright Office.

United States Copyright Office. 2025b. *Copyright and Artificial Intelligence. Part 3: Generative AI Training (Pre-Publication Version)*. United States Copyright Office.

University of Tartu. n.d. "Guidelines for Using AI Applications for Teaching and Studies | Tartu Ülikool." https://ut.ee/en/content/guidelines-using-ai-applications-teaching-and-studies.

US Environmental Protection Agency. 2024. "Greenhouse Gas Emissions from a Typical Passenger Vehicle." Overviews and Factsheets, last updated August 24, accessed September 11. https://www.epa.gov/greenvehicles/greenhouse-gas-emissions-typical-passenger-vehicle.

Uzuner, Sedef. 2008. "Multilingual Scholars' Participation in Core/Global Academic Communities: A Literature Review." *Journal of English for Academic Purposes* 7 (4): 250–63. https://doi.org/10.1016/j.jeap.2008.10.007.

Vee, Annette, Tim Laquintano, and Carly Schnitzler, eds. 2023. *TextGenEd: Teaching with Text Generation Technologies*. WAC Clearinghouse. https://doi.org/10.37514/TWR-J.2023.1.1.02.

Vettese, Troy. 2019. "Sexism in the Academy." *N+1*, May 2, 2019. https://www.nplusonemag.com/issue-34/essays/sexism-in-the-academy/.

Von Bergen, Megan. 2023. "Defining Ungrading: Alternative Writing Assessment as Jeremiad." *Composition Studies* 51 (2): 137–42.

Waigandt, Diana, Alicia Noceti, and Raquel M. T. Lothringer. 2019. "Writing for Publication in English: Some Institutional Initiatives at the Universidad Nacional de Entre Ríos." In *Pedagogies and Policies for Publishing Research in English*, edited by James N. Corcoran, Karen Englander, and Laura-Mihaela Muresan, 56–74. Milton Park, UK: Routledge.

Waldinger, Robert, and Marc Schulz. 2023. *The Good Life: Lessons from the World's Longest Scientific Study of Happiness*. New York: Simon & Schuster.

Walker, John. 2010. "Measuring Plagiarism: Researching What Students Do, Not What They Say They Do." *Studies in Higher Education* 35 (1): 41–59.

Wallwork, Adrian. 2011. *English for Writing Research Papers*. New York: Springer.

Walsh, Kat. 2023. "Understanding CC Licenses and Generative AI." Creative Commons, August 18. https://creativecommons.org/2023/08/18/understanding-cc-licenses-and-generative-ai/.

Walvoord, Barbara E. 2010. *Assessment Clear and Simple: A Practical Guide for Institutions, Departments, and General Education*. 2nd ed. Hoboken, NJ: Jossey-Bass.

Wang, Chaoran, Zixi Li, and Curtis Bonk. 2024. "Understanding Self-Directed Learning in AI-Assisted Writing: A Mixed Methods Study of Postsecondary Learners." *Computers and Education: Artificial Intelligence* 6 (June):100247. https://doi.org/10.1016/j.caeai.2024.100247.

Wargo, Katalin. 2020. "A Conceptual Framework for Authentic Writing Assignments: Academic and Everyday Meet." *Journal of Adolescent & Adult Literacy* 63 (5): 539–47.

Warner, John. 2025. *More Than Words: How to Think about Writing in the Age of AI*. New York: Basic Books.

Whitehand, J. W. R. 2005. "The Problem of Anglophone Squint." *Area* 37 (2): 228–30.

Wilder, Craig Steven. 2013. *Ebony and Ivy: Race, Slavery, and the Troubled History of America's Universities*. Illus. ed. New York: Bloomsbury Publishing.

Williams, Cheri, and Sandra Beam. 2019. "Technology and Writing: Review of Research." *Computers & Education* 128 (January): 227–42. https://doi.org/10.1016/j.compedu.2018.09.024.

Williams, James D., and Seiji Takaku. 2011. "Help Seeking, Self-Efficacy, and Writing Performance among College Students." *Journal of Writing Research* 3 (1): 1–18. https://doi.org/10.17239/jowr-2011.03.01.1.

Wolfram, Stephen. 2023. "What Is ChatGPT Doing . . . and Why Does It Work?" *Writings* (blog), February 14, 2023. https://writings.stephenwolfram.com/2023/02/what-is-chatgpt-doing-and-why-does-it-work/.

Wuchty, Stefan, Benjamin F. Jones, and Brian Uzzi. 2007. "The Increasing Dominance of Teams in Production of Knowledge." *Science* 316 (5827): 1036–39. https://doi.org/10.1126/science.1136099.

Yancey, Kathleen Black, Liane Robertson, and Kara Taczak. 2014. *Writing across Contexts: Transfer, Composition, and Sites of Writing*. Logan: Utah State University Press.

Zhai, Xiaoming, Joseph Krajcik, and James W. Pellegrino. 2021. "On the Validity of Machine Learning-Based next Generation Science Assessments: A Validity Inferential Network." *Journal of Science Education and Technology* 30 (2): 298–312. https://doi.org/10.1007/s10956-020-09879-9.

Zhai, Xiaoming, and Ross H. Nehm. 2023. "AI and Formative Assessment: The Train Has Left the Station." *Journal of Research in Science Teaching* 60 (6): 1390–98. https://doi.org/10.1002/tea.21885.

Zielinski, Chris, Margaret Winker, Rakesh Aggarwal, Lorraine Ferris, Markus Heinemann, Jose Florencio Lapeña, Sanjay Pai, Edsel Ing, and Leslie Citrome. 2023. "Chatbots, ChatGPT, and Scholarly Manuscripts—WAME Recommendations on ChatGPT and Chatbots in Relation to Scholarly Publications." *Afro-Egyptian Journal of Infectious and Endemic Diseases* 13 (1): 75–79. https://doi.org/10.21608/aeji.2023.282936.

Index